Phonons - Recent Advances, New Perspectives and Applications

Edited by Jie Deng

Published in London, United Kingdom

Phonons - Recent Advances, New Perspectives and Applications
http://dx.doi.org/10.5772/intechopen.111063
Edited by Jie Deng

Contributors
Airat Kiiamov, Alexander Rozanov, Ali Rostami, Alois Loidl, Axel Günther, Dmitrii A. Tayurskii, Dorina Croitori, Franz Mayr, Hans-Albrecht Krug von Nidda, Hodjat Ahmadi, Jie Deng, Lenar R. Tagirov, Mamoun Hemmida, Maxim Kuznetsov, Sebastian Widmann, Vladimir Tsurkan, Wenjie Guo, Zakir Seidov

First published in London, United Kingdom, 2024 by IntechOpen
IntechOpen is the global imprint of INTECHOPEN LIMITED, registered in England and Wales, registration number: 11086078, 167-169 Great Portland Street, London, W1W 5PF, United Kingdom

British Library Cataloguing-in-Publication Data
A catalogue record for this book is available from the British Library

Additional hard and PDF copies can be obtained from orders@intechopen.com

Phonons - Recent Advances, New Perspectives and Applications
Edited by Jie Deng
p. cm.
Print ISBN 978-1-83769-589-8
Online ISBN 978-1-83769-588-1
eBook (PDF) ISBN 978-1-83769-590-4

For EU product safety concerns:
IN TECH d.o.o., Prolaz Marije Krucifikse Kozulić 3, 51000 Rijeka, Croatia,
info@intechopen.com or visit our website at intechopen.com.

Meet the editor

Jie Deng is currently a researcher in the Department of Engineering, Universitat Ramon Llull, Spain, where he leads research on the vibroacoustics of acoustic black holes (ABHs), wave manipulation, energy harvesting, and acoustic metamaterials. He holds dual Ph.D.s in Mechanical Engineering from Universitat Ramon Llull (Excellent Cum Laude) and Chongqing University, China (Excellent Thesis). In 2020, he joined the Institut National des Sciences Appliquées (INSA-Lyon) as a visiting scholar. To date, Dr. Deng has authored 41 journal articles and 14 conference papers. He was listed among the top 2% of scientists in the world in 2023.

Contents

Preface

This book presents the latest developments in phonons and acoustic metamaterials. Phonons are quantum concepts used to describe lattice vibrations. In addition to phonons, the book explores phononic crystals, which are artificial materials with periodic structures, and acoustic metamaterials, which extend beyond periodicity. Recent research has delved into various aspects, including bandgaps, passband properties such as negative refraction and collimation, and topological features. New concepts have emerged in this field, including rainbow-trapping metamaterials and acoustic black holes (ABHs). These materials find applications in wave focusing, filtering, isolation, energy harvesting, particle manipulation, and topological protection, among others.

I am grateful to the authors who accepted my invitation to contribute to this book. They are Alexander Rozanov, Ali Rostami, Hodjat Ahmadi, Airat Kiiamov, Maxim Kuznetsov, Vladimir Tsurkan, Dorina Croitori, Hans-Albrecht Krug von Nidda, Zakir Seidov, Franz Mayr, Mamoun Hemmida, Sebastian Widmann, Axel Günther, Alois Loidl, Dmitrii A. Tayurskii, Lenar R. Tagirov, and Wenjie Guo.

Jie Deng
Department of Engineering,
La Salle, Universitat Ramon Llull,
Barcelona, Spain

Chapter 1

Introductory Chapter: Phonons and Acoustic Metamaterials

Jie Deng

1. Introduction

Phonons are quantum excitations in solids that describe the quantum properties of lattice vibrations. They are quantum descriptions of atomic or ionic vibrations in crystals [1]. Phononic crystals [2], on the other hand, are artificial periodic structures that manipulate the propagation of phonons through their periodicity, analogous to how electronic crystals control electrons.

Phononic crystals manipulate the propagation of phonons through periodic structures, thereby achieving control and modulation of sound waves. The design of phononic crystals can affect the phonon dispersion relations and bandgap structures, thereby influencing the propagation properties of phonons. The design and application of phononic crystals typically involve aspects such as frequency-selective phonon transmission, phononic waveguides, and phonon isolation. These applications rely on controlling the propagation of phonons. Therefore, research on phononic crystals is closely related to the properties of phonons.

Acoustic metamaterials represent a broader concept than phononic crystals. It is widely accepted that acoustic metamaterials are materials with properties that do not exist in nature. While some acoustic metamaterials still rely on periodic configurations of local resonators [3] and/or scatterers [4], or topological properties of supercells [5], their operating frequency range can be significantly lower than that of phononic crystals, especially in the sub-wavelength region. Furthermore, acoustic metamaterials can be aperiodic, achieved by artificially modifying structural or material properties. Examples include acoustic black holes (ABHs) [6], rainbow trapping [7], acoustic metasurfaces [8], and functionally graded materials [9].

2. Trends of phonons and acoustic metamaterials

Phonons and acoustic metamaterials represent a vibrant research area that has attracted substantial attention in recent years. The research trends in this field include, but are not limited to, the following:

- Exploration of nonlinear phenomena

In the realm of acoustic metamaterials, there is a burgeoning interest in investigating nonlinear phenomena. This entails scrutinizing the nonlinear behavior of acoustic metamaterials, encompassing aspects such as nonlinear wave propagation, wave

IntechOpen

mixing, and frequency conversion. By delving into nonlinear acoustic metamaterials, researchers aim to unveil novel functionalities with implications for diverse applications, including signal processing, imaging, and nonlinear acoustics.

- Advancements in active and reconfigurable architectures

Active and reconfigurable acoustic metamaterials have emerged as focal points of research endeavors. These metamaterials exhibit tunable properties that can be dynamically controlled or reconfigured in real time. Integrating actuators, sensors, or external stimuli, active acoustic metamaterials demonstrate potential for adaptive functionalities, including dynamic acoustic lenses, switches, and cloaks. Such developments hold promise for enhancing versatility and adaptability in acoustic wave manipulation.

- Pursuit of broadband and low-frequency capabilities

Efforts to achieve broadband and low-frequency performance in acoustic metamaterials continue to drive research endeavors. Broadband and low-frequency acoustic metamaterials are essential for applications such as underwater acoustic cloaking, seismic wave control, and low-frequency sound absorption. Research in this domain focuses on overcoming technical challenges to realize effective manipulation capabilities across a wide frequency spectrum, thereby expanding the applicability and utility of acoustic metamaterials.

- Exploration of topological concepts

Topological principles are increasingly being explored in the design and development of acoustic metamaterials. By harnessing topological concepts, researchers aim to achieve robust and protected sound wave propagation. Topological acoustic metamaterials may exhibit unique edge states or protected modes that are resilient to defects or disorder, offering potential applications in sound waveguiding, routing, and isolation. The pursuit of topological acoustic metamaterials represents a frontier in the quest for resilient and fault-tolerant acoustic wave manipulation.

- Advancements in acoustic black holes (ABHs)

ABHs are engineered structures designed to concentrate and absorb sound energy effectively, mimicking the way gravitational black holes trap light. They typically feature a tapered geometry that gradually slows down and compresses incoming acoustic waves, leading to their efficient absorption and dissipation. ABHs are utilized in noise reduction and vibration control applications due to their exceptional energy-trapping capabilities. Research into ABHs is focused on optimizing their design and expanding their applications in various fields, including acoustics, materials science, and engineering.

In summary, the trends mentioned above underscore the dynamic landscape of research in phonons and acoustic metamaterials. Continued exploration of nonlinear phenomena, active and reconfigurable architectures, broadband and low-frequency capabilities, topological concepts, and ABHs promises to catalyze transformative breakthroughs in acoustic wave manipulation and pave the way for novel applications across diverse scientific and technological domains.

Author details

Jie Deng
Human-Environment Research (HER) Group, Department of Engineering, La Salle, Universitat Ramon Llull C/Quatre Camins, Barcelona, Catalonia, Spain

*Address all correspondence to: dengjie_cn@outlook.com

References

[1] Srivastava GP. The Physics of Phonons. Boca Raton: CRC Press; 2022

[2] Page JH, Sukhovich A, Yang S, Cowan ML, Van Der Biest F, Tourin A, et al. Phononic crystals. Physica Status Solidi (b). 2004;**241**(15):3454-3462

[3] Liu Z, Zhang X, Mao Y, Zhu Y, Yang Z, Chan CT, et al. Locally resonant sonic materials. Science. 2000;**289**(5485):1734-1736

[4] Rybin M, Khanikaev A, Inoue M, Samusev K, Steel M, Yushin G, et al. Fano resonance between Mie and Bragg scattering in photonic crystals. Physical Review Letters. 2009;**103**(2):023901

[5] Hu B, Zhang Z, Zhang H, Zheng L, Xiong W, Yue Z, et al. Non-Hermitian topological whispering gallery. Nature. 2021;**597**(7878):655-659

[6] Pelat A, Gautier F, Conlon SC, Semperlotti F. The acoustic black hole: A review of theory and applications. Journal of Sound and Vibration. 2020;**476**:115316

[7] Alshaqaq M, Sugino C, Erturk A. Programmable rainbow trapping and band-gap enhancement via spatial group-velocity tailoring in elastic metamaterials. Physical Review Applied. 2022;**17**(2):L021003

[8] Cao L, Xu Y, Assouar B, Yang Z. Asymmetric flexural wave transmission based on dual-layer elastic gradient metasurfaces. Applied Physics Letters. 2018;**113**(18):183506

[9] Jin ZH, Batra R. Some basic fracture mechanics concepts in functionally graded materials. Journal of the Mechanics and Physics of Solids. 1996;**44**(8):1221-1235

Chapter 2

Ab initio Modeling of Phonons in the Family of Quasi-One-Dimensional Antiferromagnets $AFeX_2$

Airat Kiiamov, Maxim Kuznetsov, Vladimir Tsurkan, Dorina Croitori, Hans-Albrecht Krug von Nidda, Zakir Seidov, Franz Mayr, Mamoun Hemmida, Sebastian Widmann, Axel Günther, Alois Loidl, Dmitrii A. Tayurskii and Lenar R. Tagirov

Abstract

A review of the recent development in *ab initio* calculations of the vibrational properties of quasi-one-dimensional (1D) antiferromagnets with $AFeX_2$ structure is presented. Density functional theory (DFT + U) was applied to calculate the phonon modes specific to each element in the structure and the corresponding partial density of states (PDOS). The calculations revealed a strongly non-Debye phonon DOS. Using these results, the nuclear inelastic scattering spectra, temperature dependence of the Lamb-Mössbauer factor, infrared (IR) absorption strength, and phonon-specific heat were derived by direct summation over the phonon modes. The calculations demonstrate good agreement with the experiments and pave the way for understanding the anomalous magnetic properties of $AFeX_2$ quasi-one-dimensional antiferromagnets at a quantitative level.

Keywords: quasi-one-dimensional antiferromagnet, *ab initio* calculation of phonons, element-specific phonon density of states, phonon-specific heat, nuclear inelastic scattering spectroscopy, Mössbauer spectroscopy, magnetic susceptibility

1. Introduction

Compounds and artificial materials with reduced dimensionality attract unflagging attention because of a variety of unusual physical phenomena studied in these systems during the past several decades [1–6]. The greatest mystery is of course the high-T_c superconductivity (HTSC)—a phenomenon inextricably linked to the reduced dimensionality of the structures where it is observed [7]. An additional, but a strongly indicative feature of copper-based HTSCs, is a strong interrelation between the superconductivity and magnetism in these compounds. Indeed, HTSC can be considered as a state emerging from the parent low-dimensional antiferromagnetic order upon doping by charge carriers. Even more striking relevance and interplay between magnetism and superconductivity are observed in so-called "iron-based superconductors" —iron-containing pnictides, oxypnictides, and chalcogenides (see Refs. [8–12] and references therein). Depending on composition, stoichiometry, and substitutions, they can undergo a transition into an antiferromagnetic or superconducting ground state (if any). In the lattice structure, iron forms mostly planes with pnictogen or chalcogen ligands, so the reduced dimensionality always accompanies the iron-based superconductivity. The microscopic origin of superconductivity and the role of the crystal structure in these compounds are still under debate. This also concerns the important question about localization or delocalization of the magnetic moments of iron. Therefore, a deeper insight into the nature of superconductivity can be achieved by means of an extensive study of the materials containing building blocks similar to these systems. In this respect, we note that superconductivity is induced by pressure, even in the quasi-one-dimensional (1D) spin-ladder systems $BaFe_2X_3$ (X = S, Se) composed of FeX_4 tetrahedra building blocks as well [13–15].

In the past decades, crystallographic studies of ternary metal chalcogenides $A_xFe_yX_z$, (A = alkali metal, Tl; X = S, Se chalcogen) have revealed various distinct compositions containing tetrahedral [FeX_4] structural units, as there are A_5FeX_4, AFe_2X_2, AFe_2X_3, A_3FeX_3, $AFeX_2$, and $A_3Fe_2S_4$ [16–23]. In particular, the two-dimensional (2D) superconductors $K_xFe_{2-y}Se_2$ and (Tl, K)$Fe_{2-x}Se_2$ exhibit layers of FePn or FeCh (Pn = pnictogen, Ch = chalcogen) tetrahedra as common structural units [24–28].

The discovery of the superconductivity under high pressure in the ladder compounds $BaFe_2S_3$ and $BaFe_2Se_3$ [13–15] opens a new area of research among the iron-based superconductors' family. Iron-based chalcogenides AFe_2X_3 (A = K, Rb, Cs or Ba, and X = chalcogen) have been attracting attention due to their unique quasi-one-dimensional crystal structure and magnetism [13, 14, 29–44]. In this family, FeX_4 tetrahedra share their edges and form two-leg ladders of Fe sites. In other words, the dominant crystal structure comprises pairs of chains, "legs," with the interchain bonds of a strength similar to that along the legs, dubbed the "rungs." Note that the magnetic structure of $BaFe_2Se_3$ [31, 33] is a 1D analog to the block magnetism in $A_2Fe_4Se_5$ [45]: four Fe spins in the two-leg ladder form a ferromagnetic block while neighboring blocks are coupled antiferromagnetically. In contrast to $BaFe_2Se_3$, magnetic structures of $BaFe_2S_3$ and AFe_2Se_3 (A = K, Rb, Cs) [13, 32, 34, 35] are of the stripe type, in which the magnetic moments couple ferromagnetically along the rung and antiferromagnetically along the leg. It is remarkable that this family of materials is not layered, like all other iron-based superconductors, but instead has the geometry of a two-leg ladder. It has been shown experimentally that $BaFe_2S_3$ becomes superconducting at pressures above 10 GPa and with the highest critical temperature of T_c = 24 K [14].

In addition to the already mentioned AFe_2X_3 (A = Ba, Rb, K, Cs; X = S, Se) compounds, there is a group of iron-chalcogenide materials $AFeX_2$ and $A_3Fe_2S_4$ (A = K, Rb, Cs, Tl; X = S, Se) which also possess 1D structures, though in this case simple chains as opposed to ladders. For example, Na_5FeS_4, A_3FeX_3 (A = Na, Cs; X = S, Se), $AFeX_2$ (A = K, Cs; X = S, Se), and $A_3Fe_2S_4$ are built from discrete tetrahedral $[FeS_4]^{5-}$complexes [17], edge-sharing double-tetrahedral $[Fe_2X_6]^{6-}$complexes [20, 46, 47], or 1D linear and zigzag ${}_{\infty}{}^{1}[FeX_{4/2}]^{-}$chains [20, 48, 49], respectively. The evaluation of magnetic susceptibility data yielded a systematic reduction in the localized iron-spin value from $S = 5/2$ of discrete tetrahedra via $S = 3/2$ in double-tetrahedral complexes and down to $S = 1/2$ in linear-chain compounds [17, 20, 46–50].

In the present chapter, we focus on the quasi-1D, linear-chain, ternary iron chalcogenides $AFeX_2$ (A = K, Rb, Tl; X = S, Se) [51–54]. The crystal structure is shown in **Figure 1**. The common motif for all of these compounds is the presence of tetrahedral $[FeX_4]$ structural units, arranged edge-sharing into linear chains along the c direction and separated by the A atoms. Although superconductivity was not observed in these linear-chain compounds so far, they can be regarded as paradigm model systems, because the short Fe-Fe distance within the chains drives strong covalence effects. As a consequence, spin reduction and delocalization of the charge carriers are expected at the verge of 1D metallic conductivity. All the experimental results presented below have been measured on high-quality single crystals, the synthesis of which has been mastered in Augsburg and Kishinev by Dr. Tsurkan's research group.

Crystallographic and magnetic data of $AFeX_2$ (A = K, Rb, Tl; X = S, Se) are summarized in **Table 1**. With respect to their magnetic properties, these iron chalcogenides can be divided into the following two groups:

1. monoclinic $TlFeS_2$, $TlFeSe_2$, $KFeSe_2$, and $RbFeSe_2$ with the magnetic moments ordered perpendicular to the chains; and

2. monoclinic $KFeS_2$, $RbFeS_2$ with the magnetic moments slightly tilted from the chain axis in the ordered state.

In all these compounds, iron is formally trivalent with $3d^5$ electronic configuration. However, the ordered magnetic moment is found to be far below the formal ionic high-spin value of $5\mu_B$, which indicates significant $3d$ delocalization. This has been explained by the intimate Fe-Fe contact due to the small intra-chain Fe-Fe distance, which is not much larger than the Fe-Fe distance (2.48 Å) in metallic iron. Therefore, along the chains a certain degree of itinerancy and even 1D conductivity can be expected.

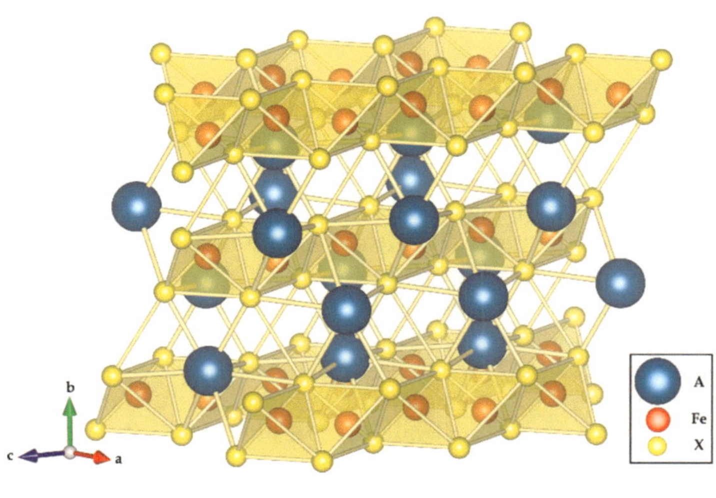

Figure 1.
Crystal structure of $AFeX_2$. The FeX_4 tetrahedra, with Fe (red sphere) in the center and X (yellow sphere) at the corners, are highlighted in a transparent yellow color. Blue large spheres denote A (adapted from Ref. [55]).

Compound	Space group	*d* (Å) (Fe-Fe)	T_N (K)	μ_{ord} (μ_B)	Moment orientation	Reference
$KFeS_2$	C2/c	2.70 uniform	250	2.43	13° from chain	[20, 48, 50, 56, 57]
$RbFeS_2$	C2/c	2.71 uniform	188	1.83	‖ chain and slightly tilted	[20, 48]
$TlFeS_2$	C2/m	2.65 dimerized	196	1.85	⊥ chain	[21, 57–59]
$KFeSe_2$	C2/c	2.81 uniform	310	3	⊥ chain	[20, 48]
$RbFeSe_2$	C2/c	2.83 uniform	250	2.66	⊥ chain	[20, 48, 55]
$TlFeSe_2$	C2/m	2.74 uniform	290	2.1	⊥ chain	[58, 60]

Table 1.
Crystallographic and magnetic data for the $AFeX_2$ (A = K, Rb, Tl; X = S, Se) (adapted from Ref. [55]).

Concerning the spin susceptibilities, Faraday magnetic balance and SQUID (superconducting quantum interference device) measurements of K, Rb, and Tl compounds $AFeX_2$ (A = K, Rb, Tl; X = S, Se) revealed the magnetic susceptibility that increases with raising the temperature (to a magnitude between 4×10^{-4} emu/mol and 2×10^{-4} emu/mol at room temperature) for rubidium (Rb) and thallium (Tl) compounds [55, 58] and passing through the maximum for the potassium compound [50]. The behavior of the susceptibility with temperature maximum is typical for one-dimensional (1D) antiferromagnetic (AFM) Heisenberg chains [61–64], with T_{max} = 565 K for $KFeS_2$, whereas $RbFeSe_2$ and $TlFeX_2$ (X = S, Se) [55, 58] compounds exhibit a continuous linear increase (up to 720 K, 550 K, and 500 K for $RbFeSe_2$, $TlFeS_2$, and $TlFeSe_2$, respectively) and do not show any tendency for saturation. This kind of a linear increase of the susceptibility is rather unusual for 1D antiferromagnetic (AFM) Heisenberg chains of localized spins, which typically exhibit a susceptibility maximum at a temperature comparable to the intra-chain exchange [61–64]. Such susceptibility maxima had been approximated using the model of an AFM spin *S* = 1/2 chain [50]. On the other hand, for $RbFeSe_2$, $TlFeS_2$, and $TlFeSe_2$, the quasi-linear susceptibility increase above T_N without any tendency to saturate [55, 58] can hardly be described within the spin-chain model. It seems that in these compounds with the strong direct Fe-Fe exchange, a fraction of *d*-electrons is close to delocalization.

The specific heat $C_P(T)$ has been measured in several works [33, 34, 37, 65–68] in the temperature range from liquid helium (L_{He}) up to room temperature (RT). Measurements reveal a weak lambda anomaly at T_N, sometimes very tiny if visible at all. As a first approach, the temperature dependence of $C_P(T)$ was fitted within classical phenomenological models including Debye and Einstein terms. The difference between the measured and the fitted $C_P(T)$ was considered as the magnetic contribution to the specific heat $C_M(T)$. The entropy changes associated with the magnetic ordering transition were estimated from the area under the lambda anomaly in C_P/T but have shown unrealistically small values, typically below 1 JK^{-1} mol^{-1}. At the antiferromagnet-paramagnet (full order-disorder) transition, the expected entropy change is close to $R\ln2$ = 5.76 JK^{-1} mol^{-1} if the iron ion is in the low-spin *S* = 1/2 state, which is an order of magnitude larger than the one obtained by the routine described above. Some authors, anyway, conclude that the small entropy change at the magnetic transitions is an indication of the low $S = 1/2$ spin state of the iron ion in $AFeX_2$ ternary chalcogenides.

The close connection between magnetic and superconducting properties in widely studied iron-containing superconductors does not allow one to study these phenomena separately [69], therefore, the study of the magnetic subsystem of the low-dimensional iron compounds is an actual problem. Two of the important quantities characterizing magnetic systems are the magnetic heat capacity and magnetic entropy. Determination of these quantities and their temperature dependences is necessary for constructing theoretical models of interactions, magnetic excitations, and superconductivity in the system.

Summarizing the introduction into the subject matter of $AFeX_2$ compounds, we note below a list of contradictions that motivated our studies:

- the magnetic moment per iron ion in the range of 1.83–3.0 μ_B is considerably reduced with respect to the expected one from ionic Fe^{3+} spin $S = 5/2$ configuration ($\sim$5.0 μ_B). At the same time, the low-spin $S = 1/2$ state by no means fits the observed magnitude of the magnetic moment;

- the values of the hyperfine field at the Fe nuclei obtained by Mössbauer spectroscopy measurements are also reduced and do not fit either the high-spin or the low-spin state of the iron ion;

- the increase of the magnetic susceptibility with raising the temperature is unusual (the Curie-Weiss descending behavior could be expected), however, for quasi-1D spin chains the Bonner-Fischer model predicts an increase of the susceptibility in a certain temperature range starting from the origin. At the moment, it is the $S = 1/2$ spin-chain model that can be applied to the particular case with great care, maybe as a hint to a possible approach to resolve the problem;

- the entropy change at T_N, estimated by phenomenological approaches, is below the lowest limit for $S = 1/2$. This indicates the insufficiency of the standard treatment of the temperature dependence of specific heat and the presence of strong fluctuations in the spin system within the entire temperature range because of the quasi-1D character of these compounds;

- *ab initio* calculations were not applied to $AFeX_2$ ternary chalcogenides before. Nowadays, it is a powerful instrument to study all kinds of physical properties of matter with reduced dimensionality.

2. Density functional theory approach for the phonon density of states

Density Functional Theory (DFT) calculations are a powerful tool to calculate the phonon density of states. This is achieved by combining DFT with phonon calculations, which are typically performed using the so-called supercell approach. This method allows for the theoretical prediction of phonon frequencies at various points in the Brillouin zone, from which the phonon density of states (PDOS) can be derived. Moreover, the PDOS provides comprehensive important information about the vibrational states of a material, which is crucial for understanding various physical properties, such as absorption spectra or specific heat.

We should note that, as a general rule, DFT calculations underestimate and inaccurately predict the forces between atoms due to certain limitations inherent in the approximation methods used for the exchange-correlation functional. The exchange-correlation functional is a crucial component in DFT that describes the quantum mechanical interactions among electrons, including exchange and correlation effects. Especially, in the case of compounds with electronic *d*-states, this is important. However, finding an exact representation of this functional is a significant challenge, and approximations must be made.

For instance, commonly used approximations, such as the Local Density Approximation (LDA) or the Generalized Gradient Approximation (GGA), can lead to inaccuracies in describing a system where electronic correlations play a significant role, such as in transition metal complexes or systems with strong covalent bonding. So, both of these complicating factors are native to the compounds under consideration because at least they contain iron ions.

Typically, in order to improve the accuracy of force predictions, researchers use more advanced functionals or add specifically designed corrections, such as the DFT + U methods or the use of hybrid functionals that mix DFT with Hartree-Fock exchange. These methods can significantly improve the accuracy of DFT predictions, including the forces between atoms.

DFT + U is an extension of DFT that is designed to improve the description of strongly correlated systems, particularly those containing transition metals, rare earth elements, or systems where *d* or *f* electrons play a significant role. In the cases of such materials, DFT + U can provide a more accurate description of electronic structures, magnetic properties, and the geometry of molecules and solids than standard DFT. It captures better the on-site electron-electron repulsion in these systems, which simple DFT underestimates. The *U* term in DFT + U refers to an added on-site Coulomb interaction term that corrects for the self-interaction error often seen in standard DFT calculation. The additional parameter *U* needs to be chosen appropriately for the system under study.

While it can lead to more accurate predictions for certain properties in correlated systems, the need for these empirical parameters can be seen as a drawback compared to parameter-free DFT methods. While DFT + U is a valuable tool for improving the accuracy of DFT calculations in systems with strong electron correlation effects, the choice of *U* is crucial and can significantly influence the results; thus, it requires careful calibration or selection based on experimental data.

So, we can conclude that the exact estimation of the interatomic forces and perfect representation of the crystal structure are two main necessary conditions to correct prediction of the phonon density of states of a compound.

Doubtless, the modern X-ray diffraction measurement techniques provide crystallographic structure data of high accuracy, even in the case of typical, generally available commercial equipment.

Conversely, the accuracy of DFT calculated forces between atoms might need meaningful efforts and application of front-line experimental methods. In our studies, we have used three ways for calibration of interatomic forces by using infrared (IR) spectroscopy, Mössbauer spectroscopy, and the method of nuclear inelastic scattering of synchrotron radiation. The detailed description of how to use such experimental data to calibrate the calculated phonon density of states is given below. Leaping ahead, we note that, as it is shown below, such high-end experimental approaches by turns need to be and might be improved by their combination with advanced DFT calculation methods.

All the *ab initio* DFT calculations were performed by means of the widely used Vienna *ab initio* simulation package (VASP 5.3) [70–73]. The electron-ion interactions were considered using the projector-augmented wave (PAW) method. This is a frozen-core method that employs the exact wave-functions of the valence electrons, instead of pseudo-wave functions [74]. Exchange and correlation corrections were taken into account by means of the Perdew-Burke-Ernzerhof (PBE) generalized gradient approximation (GGA) [75]. The detailed description of calculated parameters, where used, is given in the corresponding cited publications. Here, we just note that the Medea-Phonon software was used to obtain the phonon dispersion and density of states in harmonic approximation [76]. In the so-called direct approach to the lattice dynamics, the forces acting on all atoms are evaluated by a set of finite displacements of a few atoms within an otherwise perfect crystal. The phonon dispersion was calculated using the totally optimized equilibrium structure. In all calculations, the spin polarization due to the antiferromagnetic order of the compounds was taken into account, using the antiferromagnetic spin pattern previously obtained by neutron diffraction [48].

3. Experimental methods

Single crystals of all compounds under consideration were grown by the Bridgman method. The structural details of the crystals were investigated by conventional X-ray diffraction on powdered single crystals at room temperature using a STADI-P (STOE & Cie GmbH, Germany) diffractometer with CuK_{α} radiation. The data were analyzed by standard Rietveld refinement using the program FULLPROF [77]. For each compound, we could not detect any impurity phases above the background. The structural analysis confirmed the monoclinic structure for all systems and gave lattice parameters similar to those presented in previous studies [48].

The specific heat was measured by a relaxation method using a commercial physical properties measurement system PPMS-9 (Quantum Design, USA) in the temperature range $1.8 \leq T \leq 300$ K and in magnetic fields up to 90 kOe.

All the Mössbauer spectra presented below were measured with an accuracy of ± 0.1 K over the whole temperature range on a conventional constant-acceleration spectrometer (WissEl, Germany) equipped by a room-temperature rhodium-matrix cobalt-57 gamma-radiation source. Preliminarily, the spectrometer was calibrated at room temperature with an α-iron foil. For the measurements, the absorber was prepared by rubbing the crystals under consideration on a glue tape and packing them into a holder closed with thin aluminum foil. The preparation procedure was done in an inert argon-atmosphere dry box.

The nuclear inelastic scattering [78, 79] experiments were performed at the Dynamics Beamline P01 of PETRA III synchrotron (DESY, Germany) [80]. Utilizing the nuclear gamma resonance of ^{57}Fe at 14.413 keV, the measurements were carried out with an inline high-resolution monochromator providing an energy bandwidth of 0.9 meV full width at half maximum (FWHM). The samples with natural ^{57}Fe isotope content were measured at a stabilized temperature of 295 K.

Infrared (IR)-absorption spectra were obtained in the wavenumber range of 100–600 cm, which corresponds to the frequency range of 3–18 THz, by using an IFS 113v (Bruker, Germany) FTIR spectrometer.

4. Calculated and experimental results and their interlinked synergy

4.1 Element-specific phonons in $RbFeSe_2$ and their density of states. DFT approximation

The recital of our results will be given by starting with $RbFeSe_2$ and continued by the related compound $KFeS_2$. This kind of order is mainly conditioned just by the chronology of the studies and does not try to depict any sophisticated rule.

Figure 2 presents the DFT computed phonon density of states of $RbFeSe_2$. Each vibrational mode is simultaneously Raman and infrared active and belongs to one of the two irreducible representations of the C_2 point-group symmetry corresponding to the $RbFeSe_2$ structure. The vibrational modes of the selenium atoms exhibit relatively low frequencies and provide the major contribution to the frequency range of 1–3.5 THz. Iron atoms possess the highest oscillation frequencies, and their vibrations contribute a substantial part to the high-frequency range of 6.5–9.5 THz. The vibrational modes of the selenium atoms are distributed over the full frequency range up to 9.5 GHz, with maxima in the PDOS both in low- and high-frequency regimes. The Debye-type acoustic PDOS of the rubidium atoms without any optical contributions indicates that the Rb atoms oscillate almost decoupled from the iron atoms in the chain. Such behavior is typical for weakly bound Rb atoms.

As we noted above, it is known that the DFT approach usually underestimates vibrational frequencies. The suitable correction factor might be estimated by comparison of calculated vibrational frequencies and exact periodicity parameters of the

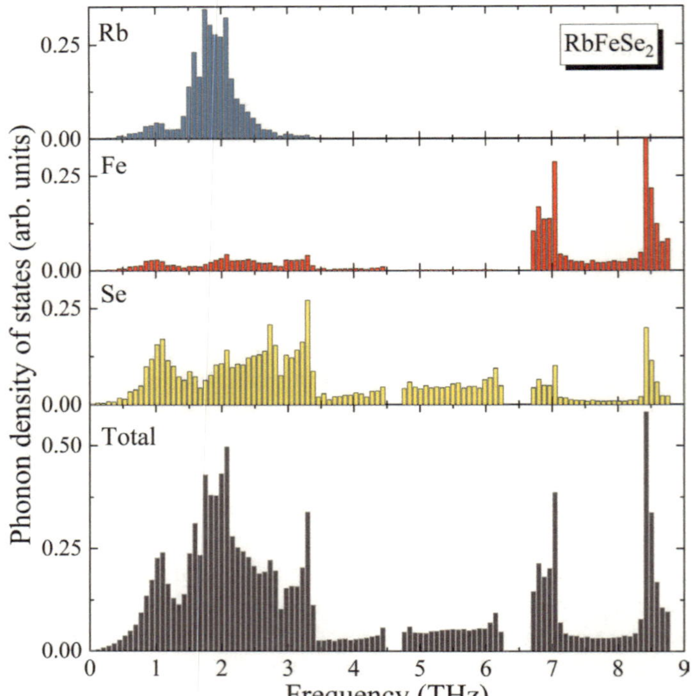

Figure 2.
Phonon density of states in $RbFeSe_2$, calculated by DFT as a function of frequency: element specific for Rb, Fe, and Se atoms (from the top toward the bottom) and in total (bottom), adapted from Ref. [51].

oscillation modes obtained by an experimental technique. In our particular case, the frequency-scale renormalization for the phonon modes has been done in three different ways:

i. by comparison of the experimentally measured temperature dependence of the Lamb-Mössbauer factor with the one calculated from the iron atoms' partial PDOS in accordance with the known relation between the Lamb-Mössbauer factor and the mean-square displacement of the resonant ion, while the last one is directly expressed via the PDOS. To accomplish that, the approximation procedure was done with the only fitting parameter, which is the oscillation frequency scale recalibration factor;

ii. by combined assay of the experimentally measured IR-absorption spectrum of the compound and the corresponding PDOS since high-frequency phonons are optically active according to the calculations;

iii. by direct comparison of calculated and experimentally measured partial PDOS of iron atoms obtained with nuclear inelastic scattering.

4.1.1 Calibration of $RbFeSe_2$ phonon frequency scale

To use the first option, we start from the initial definition of the Lamb-Mössbauer factor as the temperature dependence of the ratio of recoil-free to total nuclear resonant absorption given by [81, 82]:

$$f_M(T) = \exp\left\{-\frac{\langle x^2\rangle E_\gamma^2}{(\hbar c)^2}\right\}, \tag{1}$$

where E_γ denotes the gamma-quantum energy, c the velocity of light, and $\hbar$ the Planck constant. The value of the Lamb-Mössbauer factor directly establishes the Mössbauer spectral area for the spectrum of the corresponding iron atoms. And both of them can be considered like related by multiplier quantities.

The mean-square displacement $<x^2>$ is given by the expectation value of the squared vibrational amplitude of the iron ions along the propagation direction of the gamma radiation. Its temperature dependence can be expressed via the phonon density of states [82] as:

$$\langle x^2\rangle = \frac{\hbar}{2M}\int \frac{1}{\omega}\coth\left(\frac{\hbar\omega}{2k_BT}\right) g_{Fe}(\omega)d\omega, \tag{2}$$

with the phonon frequency ω, the partial PDOS g_{Fe} of the Mössbauer nuclei, the (effective) mass M of the Fe nucleus, and the Boltzmann constant k_B. Then, we calculate the temperature dependence of the Mössbauer spectral area, Eq. (1), by numerical integration of Eq. (2) inserting the PDOS of iron $g_{Fe}(\omega)$, given in **Figure 2**.

Figure 3 shows the adjustment of the calculated to the experimental temperature dependence of the Mössbauer spectral area. In the analysis, the oscillations of the iron ions along X, Y, and Z directions have been taken into account without weighting, because the X-ray diffraction data did not reveal any significant texture of the samples. The bare atom mass of iron has been inserted for M. The only fitting parameter

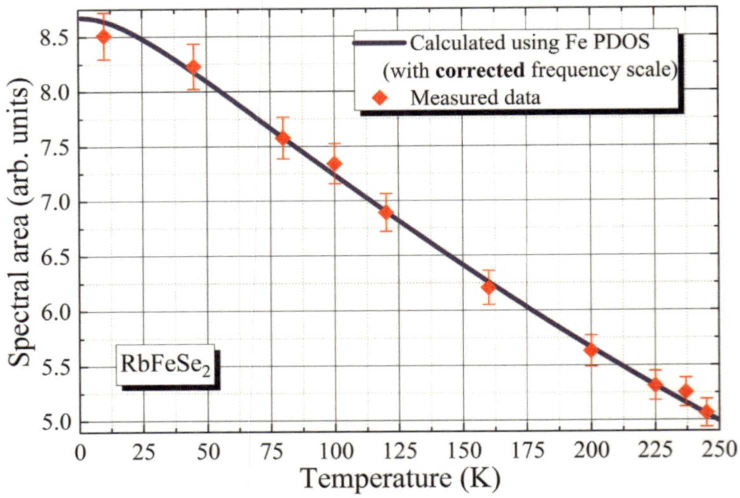

Figure 3.
Temperature dependence of the Mössbauer spectral area of $RbFeSe_2$: red solid symbols – experiment; blue solid line – calculation utilizing the iron partial PDOS (adapted from Ref. [51]).

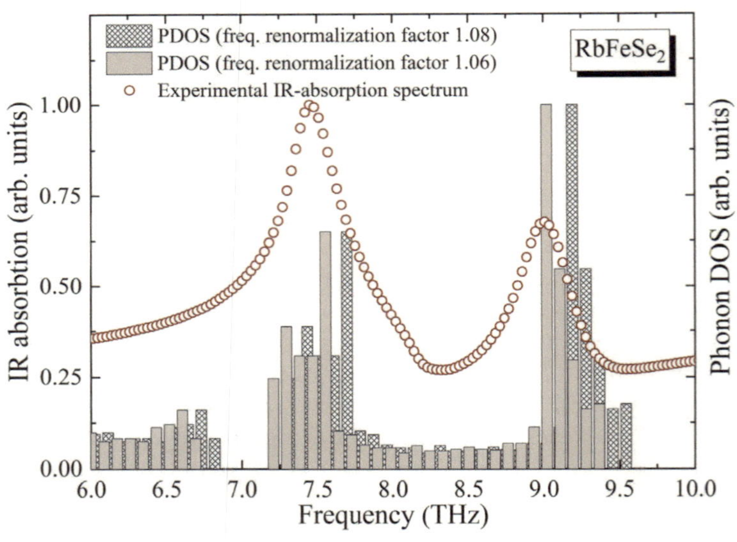

Figure 4.
Infrared-absorption spectrum of $RbFeSe_2$ compared with the calculated phonon density of states of the compound. The frequency scale of the PDOS is corrected in accordance with the Mössbauer data (dashed bar charts) and to get the best coincidence of the peaks of PDOS maxima with the IR absorption maxima (solid bar charts) (adapted from Ref. [51]).

was the scale-correction factor for the frequency as discussed above. The best agreement between the calculation and experiment has been achieved by adjusting the frequency scale by a factor of 1.08.

To do the second option we compared, as presented in **Figure 4**, the experimental IR-absorption spectrum with the calculated PDOS, where the frequency scale was corrected by the factor of 1.08, and we also performed the fitting procedure to find the best agreement between the DFT and experimental data. The IR-absorption spectrum shows the best agreement with the calculated PDOS in the case of a scaling correction factor of 1.06 (solid bar chart). The 2% difference in the scaling-factor values obtained with Mössbauer and optical experiments we qualitatively explain by possible slight crystal structural distortions that result from surface oxidation of the compound during IR-absorption spectroscopy measurements, giving rise to a corresponding phonon-frequency shift. As the IR-absorption measurements were performed in ambient air atmosphere, they should be more sensitive to modifications of the surface than the Mössbauer measurements, which were conducted under protective argon atmosphere, both the preparation of the absorber and the measurements. Nevertheless, such a proximity of these two scaling-factor values anyway appears to make these results quite persuading.

Now, we begin to consider the results obtained with using the high-end experimental technique based on the nuclear inelastic scattering (NIS) of synchrotron radiation on resonant iron nuclei. The Fe nuclear inelastic scattering spectrum of $RbFeSe_2$ is shown in **Figure 5** (blue solid hexagons) along with the instrumental function measured simultaneously with the spectrum (green solid hexagons in **Figure 5**).

The partial iron PDOS of $RbFeSe_2$ was determined from the NIS spectrum following the procedure described in Ref. [83] and is depicted in **Figure 6** (blue open hexagons). The magenta solid line in **Figure 6** shows the DFT calculated partial PDOS for the iron atoms in $RbFeSe_2$ recalibrated by using the correction factor of 1.08. The PDOS calculated by DFT consists of a discrete set of contributions from phonon modes, while the NIS-derived PDOS is a quasi-continuous sequence due to the settings of the beamline setup, but mainly due to the energy resolution of the monochromator, which amounts to 0.9 meV. To compare both with each other, the DFT calculated partial PDOS of the iron atoms was convoluted with a Gaussian profile of 0.9 meV FWHM, simulating the resolution of the monochromator. The result of the convolution was finally recalibrated as mentioned above yielding the solid magenta line presented in **Figure 6**.

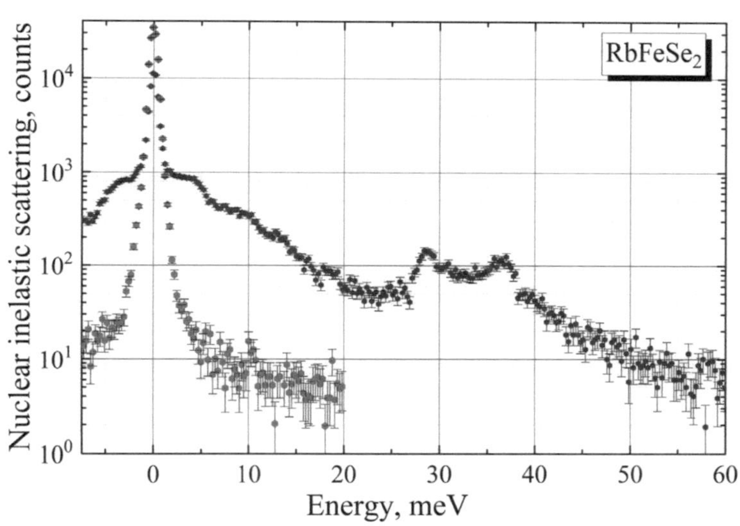

Figure 5.
^{57}Fe nuclear inelastic scattering spectrum of $RbFeSe_2$ (blue hexagons) and instrumental function (green hexagons) (adapted from Ref. [52]).

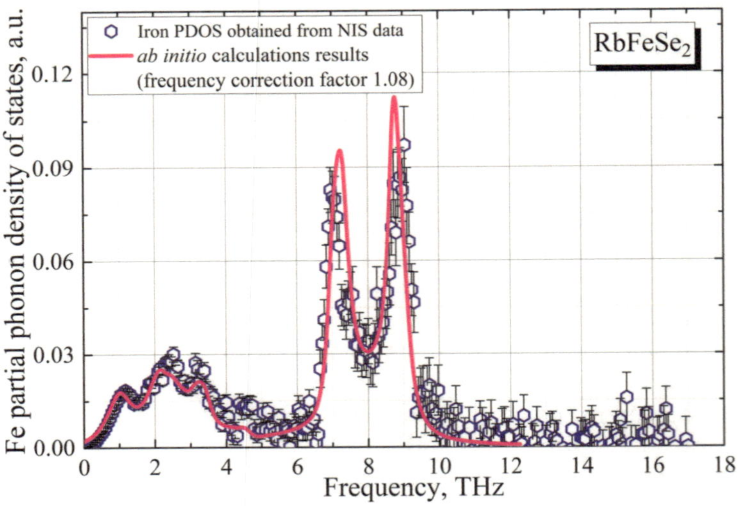

Figure 6.
Partial phonon density of states for iron atoms of $RbFeSe_2$ obtained from the nuclear inelastic scattering spectrum (blue open hexagons) and from ab initio *calculations (magenta line, see description in the body text) (adapted from Ref. [52]).*

Thus, **Figure 6** proves excellent agreement between the *ab initio* calculated PDOS and the NIS-derived PDOS. In the entire frequency range of lattice vibrations, all features of the iron PDOS in $RbFeSe_2$ are quantitatively described by the calculation. The optimal frequency correction factor of 1.077 ± 0.002 is practically identical to that obtained previously from the analysis of the temperature-dependent Lamb-Mössbauer factor. Such accordance corroborates our previous results and ensures that the temperature dependence of the Lamb-Mössbauer factor enables an accurate calibration of the DFT calculated PDOS of solids.

So, three independent experimental techniques done on different sample pieces, performed with various equipment, including both the commercial one and the unique high-end measuring setup in DESY Centre, gave almost identical results. We also note that, the last comparison of DFT results and data obtained by nuclear inelastic scattering experiments prove the excellent quantitative agreement between them. It is clearly seen that all the features of the iron PDOS are well reproduced within the entire frequency range of vibrations in the $RbFeSe_2$ lattice. So, for further quantitative analysis of the thermodynamic properties, we use the frequency scaling factor of 1.08.

As an optional side-conclusion, we need to pay attention to the fact that the approximation of the temperature dependence of the Lamb-Mössbauer factor by using the calculated PDOS and the frequency correction factor as the only fitting parameter gave the same frequency scale calibration correction as was obtained by using the high-cost and barely available nuclear inelastic scattering method based on synchrotron radiation. Despite the own difficulties of using the conventional Mössbauer spectroscopy related to operating gamma-radiation sources, this method, beyond doubt, is more accessible than nuclear inelastic scattering measurements depending on synchrotron radiation sources. Thus, both techniques can be equally applied for evaluation of the frequency correction factor for the DFT calculated phonon frequencies.

4.1.2 Temperature dependence of $RbFeSe_2$ specific heat and its analysis

In the case of $RbFeSe_2$, the total specific heat is expected to originate from two contributions, (i) the lattice contribution due to acoustic and optical phonons as discussed above and (ii) an additional magnetic contribution determined by the thermal population of excited magnetic states. The phonon density of states enables to directly calculate the lattice contribution to the specific heat by using the harmonic approximation [84]. In the harmonic approximation, the lattice heat capacity can be determined as follows [85]:

$$C_V(T) = DNk_B \int \left[\frac{\hbar\omega/2k_BT}{\sinh(\hbar\omega/2k_BT)} \right]^2 g_T(\omega)d\omega, \tag{3}$$

where D is the number of degrees of freedom in the unit cell (3 in our case) and $g_T(\omega)$ is the total PDOS (**Figure 2**, bottom panel). Traditionally experimental works [33, 34, 37, 65–68] present the temperature dependence of the specific heat at constant pressure $C_P(T)$ because the specific heat is normally measured at ambient conditions in a wide temperature range so that the thermal expansion is not negligible.
Since the general thermodynamic approach gives the following relation:

$$C_P(T) - C_V(T) = -T\left[(\partial V/\partial T)_P\right]^2/(\partial V/\partial P)_T > 0, \tag{4}$$

therefore, the specific heat at constant pressure $C_P(T)$ is always larger than the one measured at constant volume. In the case of solids, the subtraction result between these two specific heat expressions is given by a simple relation:

$$C_P(T) - C_V(T) = -T\frac{\alpha^2 V_0^2}{dV_0/dP} = \alpha^2 B V_0 T, \tag{5}$$

where $\alpha \equiv \alpha(P)$ denotes the thermal expansion coefficient, V_0 represents the molar volume, and B is the bulk modulus. The bulk modulus of $RbFeSe_2$ has been estimated directly from the second derivative of the total energy as a function of the unit-cell volume as B = 17.1 GPa. The thermal expansion coefficient α was estimated utilizing the crystal-structure data presented for room temperature and temperature of 14 K in Ref. [48]. This allowed us to express the specific heat at constant pressure as $C_P \approx C_V + 0.008[\text{J/mol K}^2]\cdot T$. **Figure 7** compares the calculated lattice heat capacity to the experimentally obtained total specific heat. The difference between the lattice specific heat and the total one is associated with the magnetic contribution to the specific heat and presented in **Figure 8**. The corresponding magnetic entropy change can be calculated from the experimentally measured magnetic specific heat as

$$\Delta S_M = \int [C_m(T)/T]dT. \tag{6}$$

The integration from zero temperature to 290 K gives the value of ΔS_M = 6.03 $JK^{-1}\ mol^{-1}$, which should be considered as an estimation of the lower boundary, because of the temperature limit (T < 290 K) of our equipment.
Note that in quasi-one-dimensional antiferromagnets, magnetic fluctuations are expected to persist up to high temperatures half of the Néel temperature and more above T_N. The experimental entropy change has to be compared to the theoretical

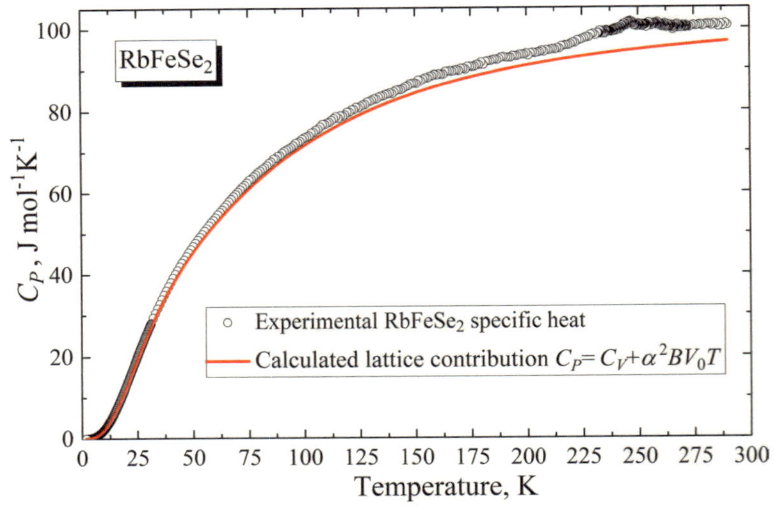

Figure 7.
Temperature dependence of the specific heat C_P *of* $RbFeSe_2$*: black circles – experimental data, red line – calculated lattice contribution to the specific heat at constant pressure* $C_P = C_V + 0.008[J/mol\ K^2]\ T$ *(adapted from Ref. [51]).*

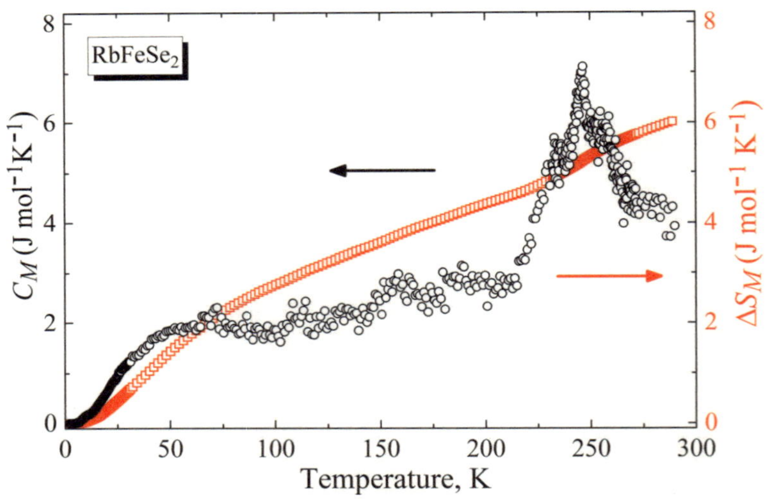

Figure 8.
Temperature dependence of the magnetic contribution to the heat capacity $C_M(T)$ *(black empty circles) and magnetic entropy change (red empty squares) of* $RbFeSe_2$ *obtained as the difference between the experimentally measured specific heat and the calculated lattice contribution (adapted from Ref. [51]).*

values ΔS_M expected at the antiferromagnetic to paramagnetic order-disorder transition for the three possibilities of low spin ½, intermediate spin 3/2, and high spin 5/2 for Fe^{3+} ($3d^5$) ions, given by $R\ln 2 = 5.76\ JK^{-1}\ mol^{-1}$, $R\ln 4 = 11.52\ JK^{-1}\ mol^{-1}$, and $R\ln 6 = 14.89\ JK^{-1}\ mol^{-1}$, respectively. Thus, for $RbFeSe_2$, the experimentally obtained value of the magnetic entropy change ΔS_M comes closest to the case of the intermediate-spin state $S = 3/2$.

In order to probe the local magnetic state of the iron ions in $RbFeSe_2$, the low-temperature Mössbauer spectroscopy data have been carefully analyzed and the hyperfine field at the ^{57}Fe nucleus was determined as $H_{0hf} = 216$ kOe. This value conforms well with Mössbauer results for the related alkali-metal and thallium-iron sulfides and selenides. Such a low value of the hyperfine field indicates a strong reduction of the local iron-spin moment. This hyperfine-field value is significantly smaller than that ($H_{0hf} \sim 500$ kOe) typical for the high-spin state of Fe^{3+} ions, but too large for the low-spin state.

Therefore, the experimentally determined hyperfine field supports the intermediate-spin state $S = 3/2$ of the iron atoms in $RbFeSe_2$ in agreement with the set of data for ΔS_M. Following Ref. [48] for edge-sharing $FeSe_4$ tetrahedra in $RbFeSe_2$, the $S = 3/2$ intermediate-spin state results from the distortion of the tetrahedra and the related splitting of the Fe^{3+} $3d$-orbital triplet and doublet by the corresponding low-symmetry crystal field. Moreover, the partial delocalization of the iron $3d$-electrons in $RbFeSe_2$ is also responsible for reduced values of hyperfine field and magnetic moment, as compared to those values typical for the $S = 3/2$ case.

The most significant magnetic contribution to the specific heat appears in the temperature range 30–200 K. Indeed, because of the quasi-one-dimensional structure of $RbFeSe_2$, one could expect that quantum fluctuations, which otherwise prevent magnetic ordering in ideal 1D spin chains according to the Mermin-Wagner theorem [86, 87], persist well below the three-dimensional (3D) ordering temperature $T_N = 248$ K. From theoretical point of view, both for 1D Heisenberg- [spin ½ to 3] or XY- [spin ½] antiferromagnetic spin chains with nearest-neighbor intra-chain exchange J, numerous papers report an approximately linear temperature dependence of the magnetic contribution to the specific heat at temperatures $T < J/k_B$. For quasi-one-dimensional antiferromagnets with interchain exchange $J' \ll J$, 3D antiferromagnetic order typically sets in at about $T_N \sim (J \times J')^{1/2}/k_B$. This means that the temperature corresponding to the intra-chain exchange energy is expected to be much larger than the Néel temperature, and therefore the temperature range covered by our measurements should undoubtedly match the regime where the quasi-linear behavior of the magnetic specific heat is predicted. However, we have to note here that usual spin-chain compounds, where existing theories are applicable, undergo magnetic order at much lower temperatures and hence, to our best knowledge, there is no appropriate theory to treat our present particular case of magnetic specific heat of quasi-one-dimensional spin-3/2 chains with the interchain interaction triggering the transition to the 3D antiferromagnetic state at a Néel temperature as high as $T_N = 248$ K.

4.2 Element-specific phonons in $KFeS_2$ and their density of states. DFT + U approximation

Further we continue with the compound $KFeS_2$, for which the same study strategy was used. **Figure 9** depicts the calculated total and element-specific PDOS as a function of frequency for $KFeS_2$ obtained by means of the same DFT approach we used above for the $RbFeSe_2$ compound. The PDOS of $KFeS_2$ is similar to that of $RbFeSe_2$, it could be divided into two distinct areas: the low- and high-frequency ones. The low-frequency area extends up to 6 THz, which shows Debye-like frequency dependence (approximately up to 3 THz), and the dominant contribution to this area comes from potassium atoms. The high-frequency area shows a significant amount of phonon density of states and consists of only vibrational modes of iron and sulfur atoms. Because of the large number of high-frequency vibrational modes of Fe and S atoms, the Debye model cannot be applied for a fine analysis of the vibrational properties of the system such as lattice specific heat. Nevertheless, this result obtained by means of

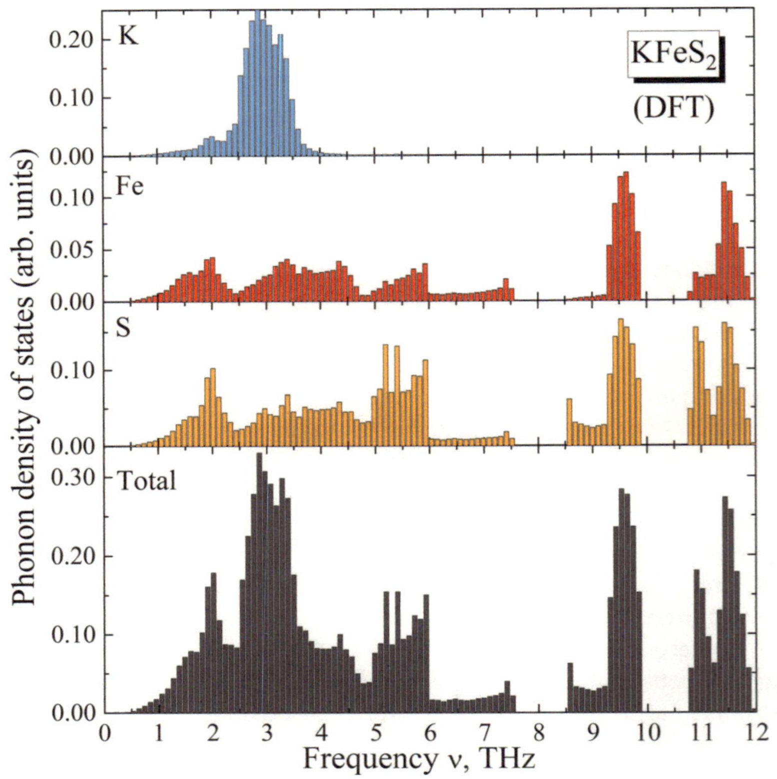

Figure 9.
The DFT calculated phonon density of states in $KFeS_2$: element-specific (K, Fe, and S atoms from the top toward the bottom) and the total PDOS (the bottom).

a simple DFT approach yielded a phonon density of states, which does not perfectly agree with the experimental results we got for $KFeS_2$. Indeed, a significant disagreement was found.

4.2.1 $KFeS_2$ PDOS and calibration of the phonon frequency scale

The above-mentioned experimental results for $KFeS_2$ are presented in **Figure 10**: (i) the partial iron phonon density of states is presented in the mainframe of **Figure 9**, and the inset gives the experimental infrared-absorption spectrum. As we noted above, the IR-absorption spectrum measures only the optical phonon modes of all three atoms in the compound, while the NIS technique probes both acoustic and optical phonon modes, but only those of the iron atoms. In the low-frequency range, the iron partial PDOS follows the cubic frequency dependence predicted by the classical Debye model for acoustic phonons. Notably, both NIS and IR measurements reveal two maxima in the optical phonon density at about 9 and 11.5 THz. Such accordance of both independent methods corroborates the high accuracy of the experimental data, while the PDOS calculated by means of the simple DFT-approach is not in accordance with them.

We should also note that the magnetic moment of iron ions in $KFeS_2$ obtained within the simple DFT-approach does not match well the experimental value obtained by neutron diffraction. This probably indicates strong delocalization of the

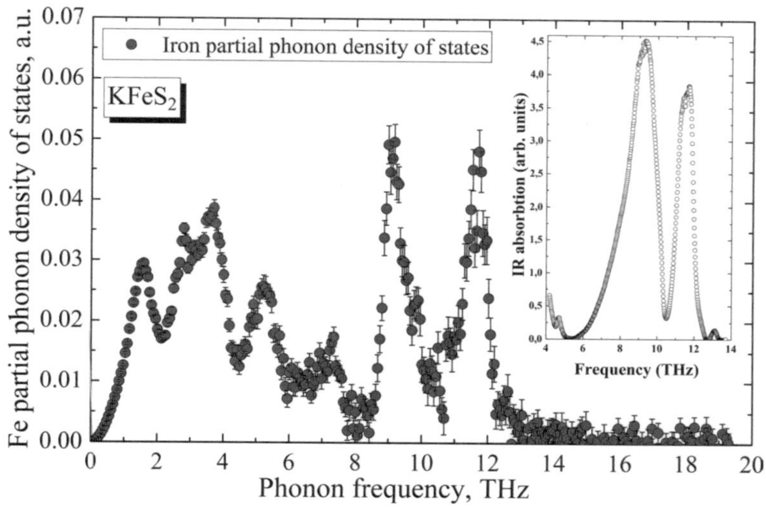

Figure 10.
Partial iron PDOS of $KFeS_2$ evaluated from the NIS spectrum of $KFeS_2$ and IR-absorption spectrum (presented in the inset) (adapted from Ref. [53]).

Fe *d*-electrons and hence a proper description of $KFeS_2$ demands to apply the DFT + U approach. The DFT + U method strongly improves the description of materials with mixed type of electron distribution. This material class includes mid-to-late first row transition metal oxides and sulfides, because the self-interaction error is largest for such multiply charged (i.e., +2 or higher) open-shell metal ions [88]. DFT + U theory does require selection of the only parameter, $U_{eff} = U - J$, the difference between intra-atomic Coulomb and exchange energies [89]. In a simplified DFT + U approach, these two parameters are not taken into account separately, but their difference is used. Sometimes the interatomic exchange energy is omitted from consideration and the only parameter U is used [90]. So, the value of the parameter U_{eff} should be preliminarily estimated from available experimental data to ensure the correct use of any DFT + U calculations of any properties of $KFeS_2$. Here, we estimate U_{eff} based on experimentally determined electronic and magnetic properties. Note that in case of $KFeS_2$, both parameters, U and J, have to be considered independently.

In case of magnetically ordered systems, the proper choice of U_{eff} is indispensable for a correct description of the magnetic ground state and calculation of the magnetic moments [91]. Thus, comparing *ab initio* results with experimental data allows to find the correct U_{eff} value just by simple selection. The magnetic ground state and ordered magnetic moment of the Fe^{3+} ions are well known from neutron diffraction measurements [55]. Below $T_N = 250$ K, $KFeS_2$ shows antiferromagnetic order with an ordered magnetic moment of 1.9 μ_B [48]. The electronic band gap below T_N was estimated from the temperature dependence of the electrical resistivity of quenched $KFeS_2$ crystals [92]. To find the proper values of the parameters U and J, we have performed a series of *ab initio* calculations of the magnetic moment and electronic band gap of $KFeS_2$ with different values of the parameters. The on-site Coulomb terms U and J have been considered independently for the strongly correlated Fe *d*-electrons. Both parameters have been adjusted to obtain a good agreement between the calculated and the experimentally determined values of the magnetic moment [48] and the band gap [92]. The best agreement was obtained for $U = 1.5$ eV and $J = 2$ eV.

The obtained *U* and *J* values allowed us to carry out the calculations of the vibrational properties of $KFeS_2$ within the DFT + U scheme. **Figure 11** shows the frequency dependence of the total and the element-specific PDOS for $KFeS_2$ calculated by this advanced way. It reveals a complex structure familiar to us, covering two distinct frequency ranges. The potassium phonon modes dominate at relatively low frequencies of 1–5 THz, while the iron atoms possess the highest vibrational frequencies and constitute a substantial part of the high-frequency range of 9–12 THz. The sulfur atoms show vibrational modes in the whole frequency range under consideration.

The comparison of the DFT + U calculated PDOS results with the experimental data of nuclear inelastic scattering measurements and the IR-absorption spectrum shows a good agreement between them. **Figure 12** proves satisfactory quantitative accordance between the iron partial PDOS derived from our NIS data and that obtained by DFT + U calculation within the entire frequency range of lattice vibrations in $KFeS_2$ after the frequency axis has been rescaled by a correction factor of 0.92.

As seen in **Figure 10**, the experimental infrared-absorption spectrum shows two broad lines. Again, like in case of the NIS data, the best agreement for the resonant frequencies of the two IR absorption peaks with the calculated ones was obtained by rescaling the frequency axis by a correction factor of 0.92.

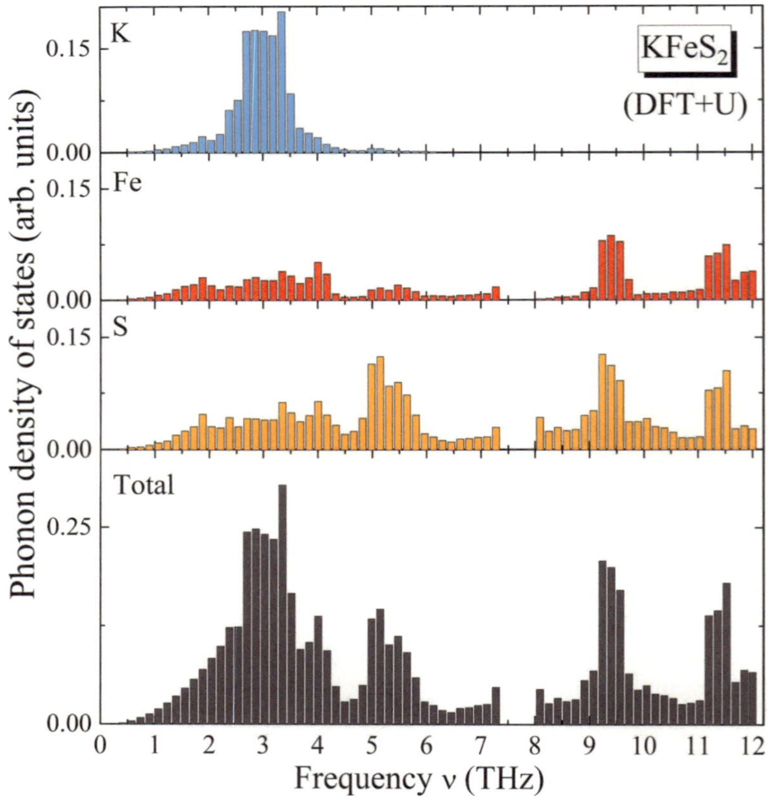

Figure 11.
Frequency dependence of the PDOS in $KFeS_2$ calculated within DFT + U approach: element-specific (K, Fe, and S atoms from the top toward the bottom) and in total (bottom). (adapted from Ref. [53]).

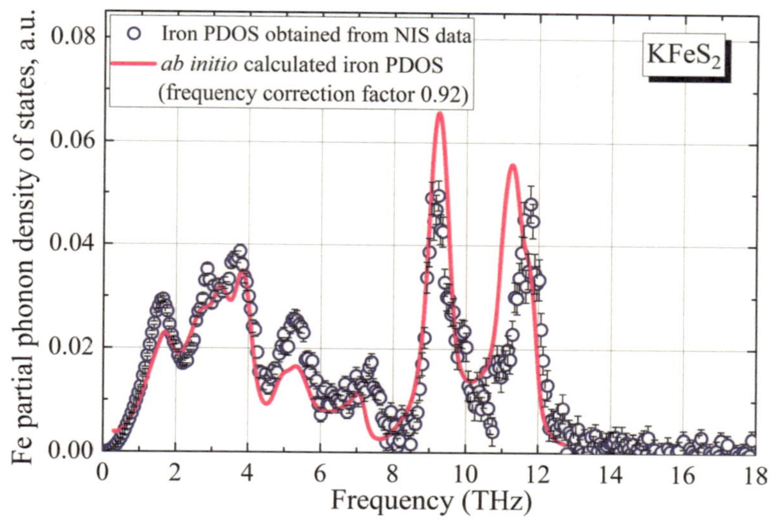

Figure 12.
Frequency dependence of the iron partial PDOS of $KFeS_2$, as derived from the NIS spectrum (blue circles), and from ab initio *calculations (magenta line), as described in the text (adapted from Ref. [53]).*

4.2.2 Temperature dependence of $KFeS_2$ specific heat and its analysis

We calculated the lattice specific heat of $KFeS_2$ utilizing the calculation procedure we used above for $RbFeSe_2$. Thereby, DFT + U calculations were applied to estimate the bulk modulus (B = 10.1 GPa) of $KFeS_2$ from the second derivative of the total energy as a function of the unit-cell volume. The total energy per unit-cell volume itself was determined from a polynomial fit, truncated after the fourth-order term. The thermal expansion coefficient α was derived from crystal-structure data of two states –at room temperature and at 14 K (Ref. [48]), hence the relation of $C_P \approx C_V + 0.003[\text{J/mol K}^2]\cdot T$ was applied to calculate the specific heat depicted in **Figure 13** (see the legend). Below 25 K, the calculated lattice contribution reveals the cubic temperature law characteristic of the Debye model for the specific heat in solids. The magnetic specific heat of $KFeS_2$, obtained after subtraction of the calculated lattice contribution from the experimental data, is shown in **Figure 14**. The corresponding magnetic entropy change was calculated from the experimental magnetic specific heat by integrating from zero temperature up to 300 K. We got $\Delta S_M = 2.98\ \text{Jmol}^{-1}\ \text{K}^{-1}$, which is an estimation of the lower boundary, because of the temperature limit ($T \leq 300$ K) of our PPMS-9 setup, compared to the maximum temperature of thermal stability of the crystal of around 900–1000 K [50]. As already mentioned above, for quasi-one-dimensional systems, magnetic fluctuations are expected to persist, even significantly above the magnetic ordering temperature. Therefore, the experimental entropy change, which is still below the value $R\ln2 = 5.76\ \text{Jmol}^{-1}\ \text{K}^{-1}$ expected for an iron low-spin $S = 1/2$ state, suggests $S = 1/2$ as the most probable spin state, but does not exclude $S = 3/2$.

Nevertheless, such verification of the accurateness of the model calculations on one hand, and obtaining information on the iron-spin state, yielding the observed magnetic moment, on the other hand, motivates further development of advanced theoretical spin-chain models beyond the well-known Bonner-Fischer model of $S = 1/2$ linear magnetic chains with anisotropic coupling [61].

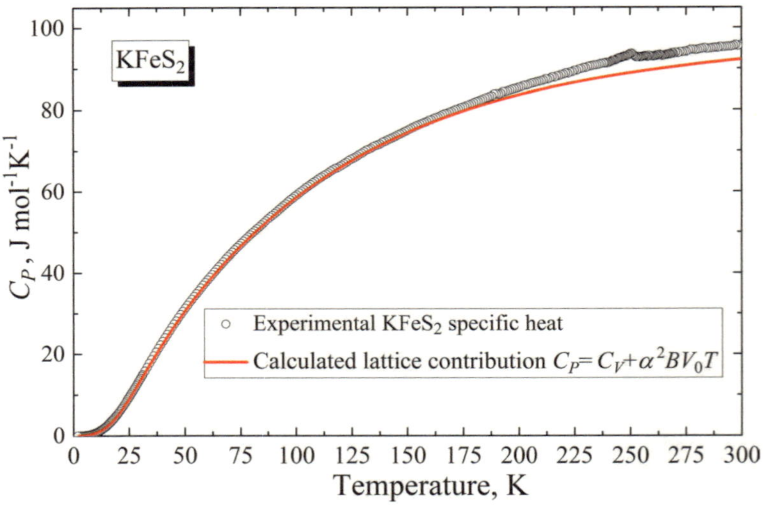

Figure 13.
Temperature dependence of the specific heat C_P *of* $KFeS_2$ *(black circles). The red line indicates the calculated lattice contribution to the specific heat at constant pressure* $C_P = C_V + 0.003[J/mol\ K^2]T$ *(adapted from Ref. [53]).*

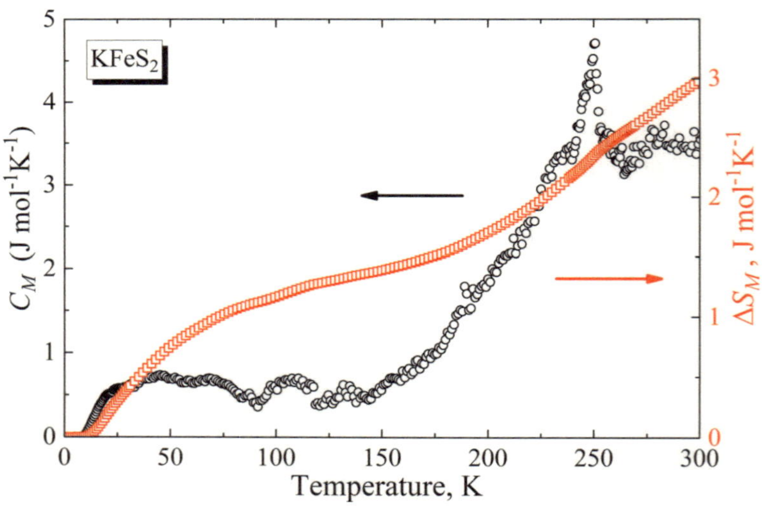

Figure 14.
Temperature dependence of the magnetic heat capacity (black empty circles), obtained as the difference between the experimental heat capacity and the calculated lattice contribution, and corresponding change of the magnetic entropy (red empty squares) of $KFeS_2$ *(adapted from Ref. [53]).*

As another optional important side-conclusion, we need to pay attention that our approach consistently describes the physical properties of quasi-one-dimensional $KFeS_2$ within DFT + U approximation. Indeed, it is well known that GGA+U functionals may underestimate band gaps, therefore, hybrid functionals [93–96] or self-interaction-corrected DFT [97, 98] techniques have been used to obtain more accurate results. We used a strategy of finding values of U and J (U = 1.5 eV and J = 2 eV [99]) parameters within the DFT + U route, which provides the total agreement with the experimentally measured magnetic moment per iron ion (1.9 μ_B) and band gap (0.35 eV). Then, the phonon frequencies were calculated from the eigenvalues of the dynamic matrix. The latter involves the evaluation of interatomic forces in all directions acting on atoms of 18 supercells for the particular $KFeS_2$ lattice. Thus, for each supercell containing 96 atoms, the dynamic matrix accumulates $96 \times 18 \times 3 = 5184$ force constants. The resulting iron partial PDOS is compared to the experimental PDOS obtained from nuclear inelastic scattering. The quite good agreement opens prospects for the analysis of thermal properties. The overall consistency of the calculated magnetic moment, electronic band gap, and iron PDOS with the experimental results in $KFeS_2$ makes the DFT + U scheme promising for studying the physical properties of related chain compounds by the low-cost DFT calculation methods. This is of special interest in the light of recent attempts to search for high-temperature superconductivity in such kind of compounds [98].

Now, we should clarify that our studies in the field of quasi-one-dimensional iron chalcogenides presented in this chapter might be applied to other compounds of that family and some results, including complex DFT + U calculations, IR-absorption spectroscopy measurements, Mössbauer spectroscopy, etc., have already been achieved. Nevertheless, in continuation of such studies on further compounds of this family, we inevitably shall have to face with the whole-aspect interpretation challenges, as we have already encountered with two considered compounds. So, despite our best efforts and the application of a wide arsenal of investigation methods and the strategy of synergetic combination of their results, it has become apparent that we are at a crossroads with no proper roadmap to choose the exact way for correct theoretical description of the behavior of iron-spin chains in the systems under consideration. Given the complexity and unique nature of the compounds we had focused on, it was imperative to use new avenues for solving the problems.

So, these innovative, modern, and unconventional approaches we used provided everyone with a wide and strong experimental facts-based fundament for further building a strong and full theoretical description of such compounds. We believe that the fresh breakthrough insights and the application of approaches based on complex theoretical physics methods will be able to overcome the obstacles regarding the full description. The potential benefit of leveraging theoretical physics approaches to devise novel solutions is obvious. We have reached a few milestones during solving the problem of a full-sided description of the family of quasi-one-dimensional chain magnets, but it needs to take a break due to the fact that the absence of key theoretical models became like a barrier to solving the problem. However, the complexity of these compounds allows considering it like model-systems for the verification of the new theoretical models which might be further generalized for the class of such complex magnets.

5. Conclusions

The lattice vibrations of ternary quasi-one-dimensional iron chalcogenides $AFeX_2$ (A = K, Rb; X = S, Se) were modeled using a unified approach based on density functional theory, taking into account intra-atomic and exchange Coulomb correlations in the d-electron system (DFT + U approach).

The phonon frequencies were calculated from the eigenvalues of the lattice dynamic matrix. The equilibrium crystal structure optimized with respect to atomic positions, cell shape, and cell volume was used for calculations of the phonon dispersion in harmonic approximation. In the antiferromagnetic phase of the $AFeX_2$ compounds, the calculations account for the spin polarization in the magnetically ordered state. It is important to note that the MedeA Phonon commercial software was allowed to perform element-specific calculations, thus providing the partial PDOS for every kind of ion in a particular $AFeX_2$ compound. It appeared that the PDOS, both partial and total, exhibit strongly non-Debye shape: vibration frequencies of cross-linking alkaline ions lie in the low-frequency domain of 2–4 THz and can be more or less satisfactorily approximated by a PDOS of the Debye type. However, the partial PDOS of iron and chalcogen vibrations represent broad bands of low-frequency contributions of strongly non-Debye shape supplemented by high-frequency, double- or triple-peak Einstein-like vibration bands in the frequency range of 6.7–12 THz. The strongly non-Debye shape of the total PDOS is a consequence of the extremely anisotropic lattice of the $AFeX_2$ family compounds and the complex structure of the elementary cell comprising rigid and soft bonds.

It is not surprising that attempts to build a phenomenological description of the temperature behavior of the heat capacity based on the Debye model led to unsatisfactory results. The addition of Einstein modes, especially their number, for better agreement with experiment was impossible to justify phenomenologically. Good agreement with experiment was found for the temperature behavior of the heat capacity in a model combining one Debye contribution and two Einstein ones, which, however, ran into a contradiction when comparing the change in magnetic entropy during the magnetic disorder/order transition. Estimates from the heat capacity measurements have given approximately 10% of $R\ln 2$ = 5.76 JK^{-1} mol^{-1}—the minimum possible change for spin S = 1/2 at an iron ion. Quite a ridiculous conclusion! Moreover, estimates of the Debye temperature obtained from measurements of the heat capacity and temperature dependence of the Lamb-Mössbauer factor differ by a factor of 2, which is also annoying to the eye. Thus, the application of the results for DFT + U phonon calculations to experiments on heat capacity could resolve the obvious contradictions that the application of phenomenological models leads to.

Before turning to the experiment and $C(T)$ analysis, we noticed that DFT approach usually underestimates frequencies of the lattice vibrations. Fortunately, nuclear inelastic scattering (NIS) technique is a direct tool to obtain the PDOS from measurements, which the calculated PDOS can be compared to, so that, the frequency correction factor could be deduced. It is an advantage of our calculations that they give element-specific PDOS, since NIS is determined solely by iron ion vibrations.

High-quality single-crystalline $AFeX_2$ samples were grown by means of the Bridgman method. The temperature dependence of the specific heat was measured. Then, to establish the frequency correction factor, nuclear inelastic scattering measurements were carried out and the partial Fe-ion PDOS was derived from the NIS spectrum. To calibrate the frequency scale of the DFT + U calculations, the calculated iron PDOS was directly adjusted to the experimental one. Regarding, for example, $RbFeSe_2$, a correction factor f = 1.08 was obtained from the best fit for $RbFeSe_2$ compound. The comparison of the experimental temperature dependence of the Lamb-Mössbauer factor with the calculated one using direct summation over DFT + U phonon modes for iron also gave the correction factor $f \simeq 1.08$. Moreover, the IR absorption measurements showed the presence of two high-frequency peaks semi-quantitatively matching the calculated PDOS after correction of frequency scale by a very similar factor f = 1.06. For DFT + U partial PDOS calculations, the correction of the frequency scale has indisputable implications on rectifying the magnetic contribution to the specific heat and the change of magnetic entropy upon the disorder-order magnetic transition. Consistency of the three experimental techniques in determining the frequency correction factor formed the basis of considering the temperature dependence of the specific heat.

After correction of the frequency scale, the *ab initio* results for the *total* PDOS were used to calculate the temperature dependence of the lattice contribution to the specific heat of $A\mathrm{FeX_2}$ without any further model approximations (neither Debye nor Einstein). Note here that also no other fitting parameters were used in the calculation. The magnetic specific heat was obtained by subtracting the calculated lattice contribution from the experimental specific heat in the temperature range of 2–300 K. The result for $C_V(T)$ was converted to $C_P(T)$ with the relationship given in Eq. (5). The resulting magnetic entropy change $\Delta S_\mathrm{M}(\mathrm{RbFeSe_2}) \simeq 6.03$ J mol^{-1} K^{-1} and $\Delta S_\mathrm{M}(\mathrm{KFeS_2}) \simeq 2.98$ J mol^{-1} K^{-1}, obtained by integrating the magnetic specific heat divided by temperature C_M/T, is close to 5.76 J mol^{-1} K^{-1}, expected for the pure low-spin ($S = 1/2$) state of the iron ion at T_N. Certainly, there is enough space to collect further entropy change up to the crystal stability temperature above 900 K. This suggests that the intermediate $S = 3/2$ spin state cannot be excluded to be realized for the iron ions in the chains. The measured magnetic moment of 2.66 μ_B and the calculated moment of $\sim$2.80 μ_B per iron ion in $\mathrm{KFeS_2}$ are in agreement with the intermediate spin state.

The estimation of the magnetic entropy change provides, therefore, a lower bound for the reduced value of the iron-spin state. This could be important for advancing the theoretical models of spin chains, generalizing the $S = 1/2$ linear magnetic chains model by Bonner-Fischer.

The approach we proposed and developed dramatically improves the consistency between the magnetic entropy change upon the disorder-order transition in quasi-one-dimensional $\mathrm{AFeX_2}$ compounds, deduced from the specific heat measurements, with the expected entropy change in the spin-chain models. However, it is a bit too early to herald that the accurate quantitative conclusion about the iron ion spin state can be made from the thermal measurements. Indeed, because the magnetic contribution to the entropy change is an integral characteristic over a wide temperature range from absolute zero up to $\sim$900 K, a limited temperature range of validity of the accepted approximations can introduce sizeable deviations from the true result. Hence, the dependence of the approximations and basic parameters on temperature must be considered, namely: harmonic vibration approximation breaks down; phonon eigenfrequencies depend on temperature, the dependence is expected to be anisotropic; the frequency correction factor can be different for particular constituent ions of the lattice; it can be inhomogeneous in frequency, etc. This tells us that there is still a lot of work to be done further to make the theory quantitative at a convincing level.

Acknowledgements

A.K., M.K., and D.A.T. acknowledge the Kazan Federal University Strategic Academic Leadership Program (PRIORITY-2030). This work was funded as joint project by the Russian Foundation for Basic Research (RFBR) and the German Research Foundation (DFG) under the contract numbers №19-52-12068 and KR2254/3-1, respectively. We are especially grateful to Hans-Christian Wille, Ilya Sergueev, Olaf Leupold, Farit Vagizov, Yury Lysogorskiy, Irina Filippova, and Tatyana Gavrilova for their invaluable contributions to original works.

Conflict of interest

The authors declare no conflict of interest.

Author details

Airat Kiiamov[1], Maxim Kuznetsov[1], Vladimir Tsurkan[2,3], Dorina Croitori[3], Hans-Albrecht Krug von Nidda[2], Zakir Seidov[2,4], Franz Mayr[2], Mamoun Hemmida[2], Sebastian Widmann[2], Axel Günther[2], Alois Loidl[2], Dmitrii A. Tayurskii[1] and Lenar R. Tagirov[5]*

1 Institute of Physics, Kazan Federal University, Kazan, Russia

2 Experimental Physics V, Center for Electronic Correlations and Magnetism, Institute of Physics, University of Augsburg, Augsburg, Germany

3 Institute of Applied Physics, Chisinau, Moldova

4 Institute of Physics, Ministry of Science and Education, Baku, Azerbaijan

5 Zavoisky Physical-Technical Institute, FRC Kazan Scientific Center of RAS, Kazan, Russia

*Address all correspondence to: ltagirov@mail.ru

References

[1] Toombs GA. Quasi-one-dimensional conductors. Physics Reports. 1978;**40**: 181-240. DOI: 10.1016/0370-1573(78) 90149-7

[2] Heimann RB, Evsyukov SE, Kavan L, editors. Physics and chemistry of materials with low-dimensional structures. Dordrecht: Springer Netherlands; 1999. 431 p. DOI: 10.1007/ 978-94-011-4742-2

[3] Morán-López JL, editor. Physics of low dimensional systems. New York: Springer US; 2001. 478 p. DOI: 10.1007/ b114803

[4] Giamarchi T. Quantum physics in one dimension. Oxford: Oxford University Press; 2003. 427 p. DOI: 10.1093/acprof: oso/9780198525004.001.0001

[5] Li Z, Wang C. One-dimensional nanostructures. Heidelberg: Springer Berlin; 2013. 146 p. DOI: 10.1007/978-3-642-36427-3

[6] Ünlü H, Horing NJM, Dabrowski J, editors. Low-dimensional and nanostructured materials and devices. New York: Springer International Publishing; 2016. 688 p. DOI: 10.1007/ 978-3-319-25340-4

[7] Eremets MI. The current status and future development of high-temperature conventional superconductivity. National Science Review. 2024;**11**: nwae047. DOI: 10.1093/nsr/ nwae047

[8] Kamihara Y, Watanabe T, Hirano M, Hosono H. Iron-based layered superconductor $La[O_{1-x}F_x]FeAs$ (x = 0.05–0.12) with T_c = 26 K. Journal of the American Chemical Society. 2008; **130**(11):3296-3297. DOI: 10.1021/ ja800073m

[9] Stewart GR. Superconductivity in iron compounds. Reviews of Modern Physics. 2011;**83**(4):1589-1652. DOI: 10.1103/RevModPhys.83.1589

[10] Fang M, Wang H, Dong C, Huang Q. Superconductivity and magnetism in (Tl, K, Rb) Fe_xSe_2. Journal of Physics: Conference Series. 2013;**449**:012015. DOI: 10.1088/1742-6596/449/1/012015

[11] Dagotto E. Colloquium: The unexpected properties of alkali metal iron selenide superconductors. Reviews of Modern Physics. 2013;**85**(2):849-867. DOI: 10.1103/RevModPhys.85.849

[12] Mancini F, Citro R, editors. The iron pnictide superconductors [Internet]. Springer Series in Solid-State Sciences. Switzerland: Springer International Publishing; 2017. DOI: 10.1007/978-3-319-56117-2

[13] Takahashi H, Sugimoto A, Nambu Y, Yamauchi T, Hirata Y, Kawakami T, et al. Pressure-induced superconductivity in the iron-based ladder material $BaFe_2S_3$. Nature Materials. 2015;**14**(10):1008-1012. DOI: 10.1038/nmat4351

[14] Yamauchi T, Hirata Y, Ueda Y, Ohgushi K. Pressure-induced Mott transition followed by a 24 K superconducting phase in $BaFe_2S_3$. Physical Review Letters. 2015;**115**(24): 246402. DOI: 10.1103/PhysRevLett. 115.246402

[15] Ying J, Lei H, Petrovic C, Xiao Y, Struzhkin VV. Interplay of magnetism and superconductivity in the compressed Fe-ladder compound $BaFe_2Se_3$. Physical Review B. 2017; **95**(24):241109. DOI: 10.1103/ PhysRevB.95.241109

[16] Bronger W. Ternary transition metal chalcogenides with framework structures and the characterization of their bonding by magnetic properties. Pure and Applied Chemistry. 1985; **57**(10):1363-1372. DOI: 10.1351/pac198557101363

[17] Klepp KO, Bronger W. Na_5FeS_4–das erste Thioferrat (III) mit diskreten tetraedrischen Anionen. Zeitschrift für Anorganische und Allgemeine Chemie. 1986;**532**(1):23-30. DOI: 10.1002/zaac.19865320105

[18] Bronger W, Ruschewitz U, Müller P. New ternary iron sulphides $A_3Fe_2S_4$ (A= K, Rb, Cs): Syntheses and crystal structures. Journal of Alloys and Compounds. 1995;**218**(1):22-27. DOI: 10.1016/0925-8388(94)01344-6

[19] Klepp KO, Sparlinek W, Boller H. Mixed valent ternary iron chalcogenides: AFe_2X_3 (A= Rb, Cs; X= Se, Te). Journal of Alloys and Compounds. 1996;**238**(1–2): 1-5. DOI: 10.1016/0925-8388(95)02087-X

[20] Bronger W, Müller P. The magnetochemical characterisation of the bonding features in ternary chalcogenides of manganese, iron and cobalt with low dimensional structural units. Journal of Alloys and Compounds. 1997;**246**(1–2):27-36. DOI: 10.1016/S0925-8388(96)02459-0

[21] Welz D, Deppe P, Schaefer W, Sabrowsky H, Rosenberg M. Magnetism of iron-sulfur tetrahedral frameworks in compounds with thallium I. Chain structures. Journal of Physics and Chemistry of Solids. 1989;**50**(3):297-308. DOI: 10.1016/0022-3697(89)90492-7

[22] Shein IR, Ivanovskii AL. Structural, electronic properties and Fermi surface of $ThCr_2Si_2$-type tetragonal KFe_2S_2, KFe_2Se_2, and KFe_2Te_2 phases as parent systems of new ternary iron-chalcogenide superconductors. Journal of Superconductivity and Novel Magnetism. 2011;**24**:2215-2211. DOI: 10.1007/s10948-011-1184-7

[23] Harrison MR, Francesconi MG. Mixed-metal one-dimensional sulfides—a class of materials with differences and similarities to oxides. Coordination Chemistry Reviews. 2011;**255**(3–4): 451-458. DOI: 10.1016/j.ccr.2010.10.008

[24] Ryu H, Lei H, Frenkel AI, Petrovic C. Local structural disorder and superconductivity in $K_xFe_{2-y}Se_2$. Physical Review B. 2012;**85**(22):224515. DOI: 10.1103/PhysRevB.85.224515

[25] Mazin II, Singh DJ, Johannes MD, Du MH. Unconventional superconductivity with a sign reversal in the order parameter of $LaFeA_sO_{1-x}F_x$. Physical Review Letters. 2008;**101**(5): 057003. DOI: 10.1103/PhysRevLett.101.057003

[26] Yan YJ, Zhang M, Wang AF, Ying JJ, Li ZY, Qin W, et al. Electronic and magnetic phase diagram in $K_xFe_{2-y}Se_2$ superconductors. Scientific Reports. 2012;**2**(1):212. DOI: 10.1038/srep00212

[27] Guo J, Jin S, Wang G, Wang S, Zhu K, Zhou T, et al. Superconductivity in the iron selenide $K_xFe_2Se_2$ ($0 \leq x \leq 1.0$). Physical Review B. 2010;**82**(18): 180520. DOI: 10.1103/PhysRevB.82.180520

[28] Fang MH, Wang HD, Dong CH, Li ZJ, Feng CM, Chen J, et al. Fe-based superconductivity with T_c= 31 K bordering an antiferromagnetic insulator in (Tl,K)Fe_xSe_2. Europhysics Letters. 2011;**94**(2):27009. DOI: 10.1209/0295-5075/94/27009

[29] Hong HY, Steinfink H. The crystal chemistry of phases in the $BaFe_2S_3$ and

Se systems. Journal of Solid State Chemistry. 1972;**5**(1):93-104. DOI: 10.1016/0022-4596(72)90015-1

[30] Krzton-Maziopa A, Pomjakushina E, Pomjakushin V, Sheptyakov D, Chernyshov D, Svitlyk V, et al. The synthesis, and crystal and magnetic structure of the iron selenide $BaFe_2Se_3$ with possible superconductivity at T_c= 11 K. Journal of Physics: Condensed Matter. 2011;**23**(40):402201. DOI: 10.1088/0953-8984/23/40/402201

[31] Caron JM, Neilson JR, Miller DC, Llobet A, McQueen TM. Iron displacements and magnetoelastic coupling in the antiferromagnetic spin-ladder compound $BaFe_2Se_3$. Physical Review B. 2011;**84**(18):180409. DOI: 10.1103/PhysRevB.84.180409

[32] Caron JM, Neilson JR, Miller DC, Arpino K, Llobet A, McQueen TM. Orbital-selective magnetism in the spin-ladder iron selenides $Ba_{1-x}K_xFe_2Se_3$. Physical Review B. 2012;**85**(18):180405. DOI: 10.1103/PhysRevB.85.180405

[33] Nambu Y, Ohgushi K, Suzuki S, Du F, Avdeev M, Uwatoko Y, et al. Block magnetism coupled with local distortion in the iron-based spin-ladder compound $BaFe_2Se_3$. Physical Review B. 2012;**85**(6): 064413. DOI: 10.1103/PhysRevB.85.064413

[34] Du F, Ohgushi K, Nambu Y, Kawakami T, Avdeev M, Hirata Y, et al. Stripelike magnetism in a mixed-valence insulating state of the Fe-based ladder compound $CsFe_2Se_3$. Physical Review B. 2012;**85**(21):214436. DOI: 10.1103/PhysRevB.85.214436

[35] Wang M, Yi M, Jin S, Jiang H, Song Y, Luo H, et al. Spin waves and magnetic exchange interactions in the spin-ladder compound $RbFe_2Se_3$. Physical Review B. 2016;**94**(4):041111. DOI: 10.1103/PhysRevB.94.041111

[36] Chi S, Uwatoko Y, Cao H, Hirata Y, Hashizume K, Aoyama T, et al. Magnetic precursor of the pressure-induced superconductivity in Fe-ladder compounds. Physical Review Letters. 2016;**117**(4):047003. DOI: 10.1103/PhysRevLett.117.047003

[37] Lei H, Ryu H, Frenkel AI, Petrovic C. Anisotropy in $BaFe_2Se_3$ single crystals with double chains of FeSe tetrahedra. Physical Review B. 2011;**84**(21):214511. DOI: 10.1103/PhysRevB.84.214511

[38] Hirata Y, Maki S, Yamaura J, Yamauchi T, Ohgushi K. Effects of stoichiometry and substitution in quasi-one-dimensional iron chalcogenide $BaFe_2S_3$. Physical Review B. 2015; **92**(20):205109. DOI: 10.1103/PhysRevB.92.205109

[39] Monney C, Uldry A, Zhou KJ, Krzton-Maziopa A, Pomjakushina E, Strocov VN, et al. Resonant inelastic x-ray scattering at the FeL_3 edge of the one-dimensional chalcogenide $BaFe_2Se_3$. Physical Review B. 2013;**88**(16):165103. DOI: 10.1103/PhysRevB.88.165103

[40] Ootsuki D, Saini NL, Du F, Hirata Y, Ohgushi K, Ueda Y, et al. Coexistence of localized and itinerant electrons in $BaFe_2X_3$ (X = S and Se) revealed by photoemission spectroscopy. Physical Review B. 2015;**91**(1):014505. DOI: 10.1103/PhysRevB.91.014505

[41] Luo Q, Nicholson A, Rincón J, Liang S, Riera J, Alvarez G, et al. Magnetic states of the two-leg-ladder alkali metal iron selenides AFe_2Se_3. Physical Review B. 2013;**87**(2):024404. DOI: 10.1103/PhysRevB.87.024404

[42] Dong S, Liu JM, Dagotto E. $BaFe_2Se_3$: A high-T_C magnetic multiferroic with

large ferrielectric polarization. Physical Review Letters. 2014;**113**(18):187204. DOI: 10.1103/PhysRevLett.113.187204

[43] Popović ZV, Šćepanović M, Lazarević N, Opačić M, Radonjić MM, Tanasković D, et al. Lattice dynamics of $BaFe_2X_3$ (X = S, Se) compounds. Physical Review B. 2015;**91**(6):064303. DOI: 10.1103/PhysRevB.91.064303

[44] Takubo K, Yokoyama Y, Wadati H, Iwasaki S, Mizokawa T, Boyko T, et al. Orbital order and fluctuations in the two-leg ladder materials $BaFe_2X_3$ (X = S and Se) and $CsFe_2Se_3$. Physical Review B. 2017;**96**(11):115157. DOI: 10.1103/PhysRevB.96.115157

[45] Ye F, Chi S, Bao W, Wang XF, Ying JJ, Chen XH, et al. Common crystalline and magnetic structure of superconducting $A_2Fe_4Se_5$ (A = K, Rb, Cs, Tl) single crystals measured using neutron diffraction. Physical Review Letters. 2011;**107**(13):137003. DOI: 10.1103/PhysRevLett.107.137003

[46] Welz D, Bennington SM, Müller P. Inverted biquadratic exchange of Heisenberg antiferromagnetic dimers in Na_3FeS_3. Physica B: Condensed Matter. 1995;**213**:339-341. DOI: 10.1016/0921-4526(95)00149-4

[47] Schmidtke HH, Rosellen U, Diehl M. Magnetic and spectroscopic high spin-low spin properties of Fe(III) sulphide complexes with extended structures. Molecular Physics. 1994;**83**(6):1191-1207. DOI: 10.1080/00268979400101871

[48] Bronger W, Kyas A, Müller P. The antiferromagnetic structures of $KFeS_2$, $RbFeS_2$, $KFeSe_2$, and $RbFeSe_2$ and the correlation between magnetic moments and crystal field calculations. Journal of Solid State Chemistry. 1987;**70**(2):262-270. DOI: 10.1016/0022-4596(87)90065-X

[49] Bronger W. Ternary sulfides: Model compounds for the correlation of crystal structure and magnetic properties. Angewandte Chemie International Edition England. 1981;**20**(1):52-62. DOI: 10.1002/anie.198100521

[50] Tiwary SK, Vasudevan S. Regular versus alternating $(FeS_4)_n$ chains: Magnetism in $KFeS_2$ and $CsFeS_2$. Physical Review B. 1997;**56**(13):7812-7814. DOI: 10.1103/PhysRevB.56.7812

[51] Kiiamov AG, Lysogorskiy YV, Vagizov F, Tagirov LR, Tayurskii DA, Seidov Z, et al. Vibrational properties and magnetic specific heat of the covalent chain antiferromagnet $RbFeSe_2$. Physical Review B. 2018;**98**(8):214411. DOI: 10.1103/PhysRevB.98.214411

[52] Kiiamov A, Tsurkan V, Croitori D, von Nidda H-AK, Seidov Z, Wille H-C, et al. Application of nuclear inelastic scattering spectroscopy to the frequency scale calibration of *ab initio* calculated phonon density of states of quasi-one-dimensional iron ternary chalcogenide $RbFeSe_2$. Applied Sciences (MDPI). 2020;**10**:7212. DOI: 10.3390/app10207212

[53] Kiiamov A, Kuznetsov M, Croitori D, Filippova I, Tsurkan V, von Nidda H-AK, et al. Density functional approach to the vibrational properties and magnetic specific heat of the covalent chain antiferromagnet $KFeS_2$. Molecules (MDPI). 2022;**27**:2663. DOI: 10.3390/molecules27092663

[54] Kiiamov A, Seidov Z, Croitori D, Tsurkan V, von Nidda H-AK, Günther A, et al. Magnetic and vibrational properties of the covalent chain antiferromagnet $RbFeS_2$. Journal of

Physics: Conference Series. 2022;**2164**: 012026. DOI: 10.1088/1742-6596/2164/1/012026

[55] Seidov Z, Krug Von Nidda HA, Tsurkan V, Filippova IG, Günther A, Gavrilova TP, et al. Magnetic properties of the covalent chain antiferromagnet $RbFeSe_2$. Physical Review B. 2016; **94**(13):134414. DOI: 10.1103/PhysRevB.94.134414

[56] Nishi M, Ito Y. Magnetic structure of $KFeS_2$—a linear chain antiferromagnet and a spin analog of active sites of two iron ferredoxins—by neutron diffraction. Solid State Communications. 1979;**30**(9):571-574. DOI: 10.1016/0038-1098(79)91138-4

[57] Welz D, Kohgi M, Endoh Y, Nishi M, Arai M. High-energy spin waves in the linear-chain antiferromagnet $KFeS_2$. Physical Review B. 1992;**45**(21): 12319-12325. DOI: 10.1103/PhysRevB.45.12319

[58] Seidov Z, Krug Von Nidda HA, Hemberger J, Loidl A, Sultanov G, Kerimova E, et al. Magnetic susceptibility and ESR study of the covalent-chain antiferromagnets $TlFeS_2$ and $TlFeSe_2$. Physical Review B. 2001; **65**(1):014433. DOI: 10.1103/PhysRevB.65.014433

[59] Welz D, Itoh S, Taylor AD. Magnetic excitations and quantum fluctuation in the covalent-chain antiferromagnet $TlFeS_2$. Europhysics Letters. 1996;**34**(4): 293. DOI: 10.1209/epl/i1996-00453-y

[60] Asgerov EB, Dang NT, Beskrovnyy AI, Madadzada AI, Ismayilov DI, Mehdiyeva RN, et al. Magnetic structure of $TlFeS_2$ and $TlFeSe_2$ chalcogenides. Semiconductors. 2015;**49**:879-882. DOI: 10.1134/S1063782615070039

[61] Bonner JC, Fisher ME. Linear magnetic chains with anisotropic coupling. Physics Review. 1964;**135**(3A): A640-A658. DOI: 10.1103/PhysRev.135.A640

[62] Bonner JC. One-dimensional model systems: Theoretical survey. Journal of Applied Physics. 1978;**49**(3):1299-1304. DOI: 10.1063/1.325026

[63] De Jongh LJ, Miedema AR. Experiments on simple magnetic model systems. Advances in Physics. 2001; **50**(8):947-1170. DOI: 10.1080/00018730110101412

[64] Hone DW, Richards PM. One- and two-dimensional magnetic systems. Annual Review of Materials Science. 1974;**4**(1):337-363. DOI: 10.1146/annurev.ms.04.080174.002005

[65] Johnston DC, Mraw SC, Jacobson AJ. Observation of the antiferromagnetic transition in the linear chain compound $KFeS_2$ by magnetic susceptibility and heat capacity measurements. Solid State Communications. 1982;**44**(2):255-258. DOI: doi.org/10.1016/0038-1098(82)90443-4

[66] Aldzhanov MA, Guseinov NG, Sultanov GD, Nadzafzade MD. Magnetic heat capacity and susceptibility of the pseudo-one-dimensional magnetic systems $TlFeS_2$ and $TlFeSe_2$. Physica Status Solidi (B). 1990;**159**(2):K107-K110. DOI: 10.1002/pssb.2221590257

[67] He JB, Wang DM, Shi HL, Yang HX, Li JQ, Chen GF. Synthesis, structure, and magnetic properties of the layered iron oxychalcogenide $Na_2Fe_2Se_2O$. Physical Review B. 2011;**84**(20):205212. DOI: 10.1103/PhysRevB.84.205212

[68] Lei H, Ryu H, Ivanovski V, Warren JB, Frenkel AI, Cekic B, et al. Structure and physical properties of the

layered iron oxychalcogenide $BaFe_2Se_2O$. Physical Review B. 2012;**86**(19):195133. DOI: 10.1103/PhysRevB.86.195133

[69] Chubukov AV, Efremov DV, Eremin I. Magnetism, superconductivity, and pairing symmetry in iron-based superconductors. Physical Review B. 2008;**78**(13):134512. DOI: 10.1103/PhysRevB.78.134512

[70] Kresse G, Hafner J. *Ab initio* molecular dynamics for liquid metals. Physical Review B. 1993;**47**(1):558-561. DOI: 10.1103/PhysRevB.47.558

[71] Kresse G, Hafner J. *Ab initio* molecular-dynamics simulation of the liquid-metal–amorphous-semiconductor transition in germanium. Physical Review B. 1994;**49**(20):14251-14269. DOI: 10.1103/PhysRevB.49.14251

[72] Kresse G, Furthmüller J. Efficiency of *ab-initio* total energy calculations for metals and semiconductors using a plane-wave basis set. Computational Materials Science. 1996;**6**(1):15-50. DOI: 10.1016/0927-0256(96)00008-0

[73] Kresse G, Furthmüller J. Efficient iterative schemes for *ab initio* total-energy calculations using a plane-wave basis set. Physical Review B. 1996; **54**(16):11169-11186. DOI: 10.1103/PhysRevB.54.11169

[74] Blöchl PE. Projector augmented-wave method. Physical Review B. 1994; **50**(24):17953-17979. DOI: 10.1103/PhysRevB.50.17953

[75] Perdew JP, Burke K, Ernzerhof M. Generalized gradient approximation made simple. Physical Review Letters. 1996;**77**(18):3865-3868. DOI: 10.1103/PhysRevLett.77.3865

[76] Parlinski K, Li ZQ, Kawazoe Y. First-principles determination of the soft mode in cubic ZrO_2. Physical Review Letters. 1997;**78**(21):4063-4066. DOI: 10.1103/PhysRevLett.78.4063

[77] Rodríguez-Carvajal J. Recent advances in magnetic structure determination by neutron powder diffraction. Physica B: Condensed Matter. 1993;**192**(1–2):55-69. DOI: 10.1016/0921-4526(93)90108-I

[78] Chumakov A, Rüffer R. Nuclear inelastic scattering. Hyperfine Interactions. 1998;**113**(1):59-79. DOI: 10.1023/A:1012659229533

[79] Rüffer R, Chumakov AI. Nuclear inelastic scattering. Hyperfine Interactions. 2000;**128**(1):255-272. DOI: 10.1023/A:1012643918108

[80] Wille HC, Franz H, Röhlsberger R, Caliebe WA, Dill FU. Nuclear resonant scattering at PETRA III: Brillant opportunities for nano–and extreme condition science. Journal of Physics: Conference Series. 2010;**217**:012008. DOI: 10.1088/1742-6596/217/1/012008

[81] Wertheim GK. Mössbauer Effect: Principles and Applications. New York and London: Academic Press; 2013. 120 p. ISBN: 9781483228563

[82] Gütlich P, Bill E, Trautwein AX. Mössbauer Spectroscopy and Transition Metal Chemistry: Fundamentals and Applications. Berlin Heidelberg: Springer-Verlag; 2010. 568 p. DOI: 10.1007/978-3-540-88428-6

[83] Kohn VG, Chumakov AI. DOS: Evaluation of phonon density of states from nuclear resonant inelastic absorption. Hyperfine Interactions. 2000;**125**(1): 205-221. DOI: 10.1023/A:1012689705503

[84] Kittel C. Introduction to Solid State Physics. New York USA: Wiley; 1996; 689 p. ISBN 0-471-11181-3

[85] Ziman JM. Electrons and Phonons: The Theory of Transport Phenomena in Solids. Oxford UK: Oxford University Press; 2001. 554 p. ISBN: 9780198507796; 0198507798

[86] Mermin ND, Wagner H. Absence of ferromagnetism or antiferromagnetism in one- or two-dimensional isotropic Heisenberg models. Physical Review Letters. 1966;**17**(22):1133-1136. DOI: 10.1103/PhysRevLett.17.1133

[87] Gelfert A, Nolting W. The absence of finite-temperature phase transitions in low-dimensional many-body models: A survey and new results. Journal of Physics: Condensed Matter. 2001;**13**(27): R505. DOI: 10.1088/0953-8984/13/27/201

[88] Bendeddouche Z, Zaoui A, Kacimi S, Abbaoui S, Kadiri A, Boukortt A. Electronic structure of transition-metal pnictide oxides $La_3T_4As_4O_2$ (T= Ni, Cu) from *ab initio* calculations. JETP Letters. 2020;**111**:210-217. DOI: 10.1134/S0021364020040013

[89] Carter EA. Challenges in modeling materials properties without experimental input. Science. 2008; **321**(5890):800-803. DOI: 10.1126/science.11580

[90] Azam S, Vu TV, Mirza DH, Irfan M, Goumri-Said S. Effect of Coulomb interactions on optoelectronic properties of Eu doped lanthanide stannates pyrochlore: DFT+U investigations. Journal of Solid State Chemistry. 2020; **290**:121522. DOI: 10.1016/j.jssc.2020.121522

[91] Rohrbach A, Hafner J, Kresse G. Electronic correlation effects in transition-metal sulfides. Journal of Physics: Condensed Matter. 2003; **15**(6):979. DOI: 10.1088/0953-8984/15/6/325

[92] Nishioka S, Kuriyaki H, Hirakawa K. Electric conduction in quasi-one-dimensional compound $KFeS_2$. Synthetic Metals. 1995;**71**(1–3):1877-1878. DOI: 10.1016/0379-6779(94)03087-M

[93] Heyd J, Scuseria GE, Ernzerhof M. Hybrid functionals based on a screened Coulomb potential. The Journal of Chemical Physics. 2003;**118**(18): 8207-8215. DOI: 10.1063/1.1564060

[94] Krukau AV, Vydrov OA, Izmaylov AF, Scuseria GE. Influence of the exchange screening parameter on the performance of screened hybrid functionals. The Journal of Chemical Physics. 2006;**125**(22):224106. DOI: 10.1063/1.2404663

[95] Qin L, Duan Y, Shi H, Shi L, Tang G. Hybrid density functional theory studies of AlN and GaN under uniaxial strain. Journal of Physics: Condensed Matter. 2012;**25**(4):045801. DOI: 10.1088/0953-8984/25/4/045801

[96] Lv D, Duan Y, Zhao B, Qin L, Shi L, Tang G, et al. Comparing the effects of uniaxial and biaxial strains on the structural stability and electronic structure in wurtzite ZnS. Journal of Applied Physics. 2013;**114**(2):023514. DOI: 10.1063/1.4813618

[97] Tang C, Ostrikov KK, Sanvito S, Du A. Prediction of room-temperature ferromagnetism and large perpendicular magnetic anisotropy in a planar hypercoordinate FeB_3 monolayer. Nanoscale Horizons. 2021;**6**(1):43-48. DOI: 10.1039/D0NH00598C

[98] Zhao X, Ma F, Lu ZY, Xiang T. $AFeSe_2$ (A= Tl, K, Rb, or Cs): Iron-based superconducting analog of the cuprates.

Physical Review B. 2020;**101**(18): 184504. DOI: 10.1103/PhysRevB.101.184504

[99] Kiiamov AG, Kuznetsov MD, Batulin RG, Tayurskii DA. On the *ab initio* DFT+U calculations of the physical properties of a compound with strong electron–electron correlations by the case of $KFeS_2$. JETP Letters. 2022; **115**(2):98-101. DOI: 10.1134/S0021364022020023

Chapter 3

Spectral-Coherency Detection of Periodic and Nonperiodic Acoustic Radiation in Discrete Rock Medium

Alexander Rozanov

Abstract

This work is performed within laboratory-field studies begun in 2015 at the mines of Khibiny appatite deposit, Kola Peninsula, Russia. The studies are aimed to determine and implement seismic criteria of risky dynamic behavior in the areas of extreme mining pressure. We observe and fix different kinds of rock dynamics – from local scale (tunnel fracturing) to regional scale (induced earthquakes). The underlying factor of such a risky dynamic behavior is the tectonics acting in directions close to horizontal. We apply the acoustic emission (AE) technique to control local instabilities in rocks. For further development of the method, we need to search for the physically-proved AE parameters that deliver information about the focal nature of hazards in mines. I suggest an approach based on spectral-correlation analysis to study a selected number of AE events. This group of signals is characterized by a two-component spectral structure and manifests itself by a specific Π-pattern coherence function. Relying on the long-term observations and considering the relation of these events to the fractured geological structure, I attribute this AE type to the *stick-slip* phenomenon. Detailed analysis of AE data revealed two natures of these AE-type sources: impulse (non-periodic) and regular (quasi-periodic) radiation.

Keywords: physical acoustics, acoustic emission, signal analysis, spectral-coherency analysis, rock mechanics

1. Introduction

Classic seismology treats the problem of determination of natural seismic source parameters through the source time function (STF) [1]. STF is one of the main parameters of the kinematic model of extended source, such as geological fault or sharp boundary, between geological formations. Once occurred on fault surfaces under mechanical stress such rock discontinuity as a shear dislocation propagates with a certain velocity along an extended source. Shear dislocation is described by the slip vector Δu corresponding to a relative displacement of the two sides of a fault. The time dependence of the slip Δu derivative is called STF. Dislocations also appear on microscale due to crystalline lattice imperfections. Their dislocations move along the slip plane as a cross line of broken atomic bonds [2, 3]. Shearing can occur on the grain boundaries of different stiffness in polycrystalline material such as rocks or on micro-fracture surfaces [4]. The dislocation cumulative process manifests itself on the macroscale as irreversible deformation, progressive development of the latter is accompanied by discrete acoustic radiation. This phenomenon is called an acoustic emission (AE).

We intend to develop our study of the AE process in rocks of different lithology in the direction of physical insight on how the strength of material relates to the defect structure evolution from micro-fractures to the fault when subjected to stress or/and temperature. To take on this problem one should think about a way to identify correctly the scale and deformation mechanism of fracturing by means of AE parameters. This is an inversion problem and it derives the properties of AE source from the acoustic far-field observations. But geophysical medium that influences the origin wave propagation is strongly inhomogeneous, thus we should find a method to account for the transfer function of the system «medium and seismic sensor (AE transducer)». In the suggested chapter, I consider the transformation process, which is associated with normal modes resulting from acoustical impact. To study and interpret a series of AE events acquired in mines of Khibiny appatite deposit in the frequency range of 2–28 kHz the spectral-correlation analysis is applied. The coherence function calculations revealed a specific spectrum pattern, which is attributed to the *stick-slip* seismic source activated by tectonic forces. In terms of deformation mode, the *stick-slip* event represents an unstable sliding on fault surfaces [1, 5, 6] characterized by two phases—*stick* as elastic energy storage, and *slip*—as energy release. Calculations of statistically significant values of spectral densities [7, 8] allowed two types of AE signals to be distinguished—non-periodic and quasi-periodic.

The experiments on fracture generation in glass plates showed that the initial acoustic radiation produced by running fracture is characterized by two parts: the compressional part and subsequent dilatational part [9, 10]. Thus, the arriving pulse can be described as a bipolar pulse. The transfer function of linear system «medium and seismic sensor» we simulate as a superposition of exponentially attenuating normal modes on the corresponding frequencies. Then, as is shown in Ref. [11], the transformation of initial pulsed radiation to the resultant AE signal we derive from convolution integral producing in an output a transient process. Here I apply the normal mode approach to the analysis of AE transducer amplitude-frequency response characteristics. Whereas the normal mode phenomenon can be observed also in the discrete geophysical medium. For example, in Ref. [12] we describe the effect of waveguide propagation of the 1st-order normal mode that arose in a horizontal ore body of thickness 30–40 m at the depth of 1000 m (Norilsk ore field, Russia). A possible theoretical view of this observation is presented in Ref. [13]. The frequency band of this wave has amounted to 40–100 Hz. In [14] normal modes of the zero order ranging from 6 to 21 Hz have been identified by means of spectral analysis. These modes correspond to the vibrations of a part of limestone massif with a volume of about 1000 m^3 in French Alps separated by an open fracture of about 15 m depth. Seismic monitoring of unstable rock volume (260,000 m^3) in the Eastern Alps, the Hochvogel summit, includes modal analysis as well aiming to control variations of rock properties due to the influence of ambient thermal factors [15]. On the other hand, the authors observe pulse-like seismic activity of 1–2 day long periods, which they presumably link to an early stage of *stick-slip* evolution of the unstable rock mass at the summit induced by mass wasting processes on the catastrophic fracture of 60 m depth.

Finally, regarding the laboratory scale of rock investigations, we evidence the normal mode phenomenon as a stop-band behavior [16] in the spectra of registered waves when the modes related to the fracture spacing stop propagating through a periodic structure of fractures [17]. Such a behavior has been observed on a frequency of about 80 kHz [16] in the core of rock salt from the Asse mine, Germany, during its critical stage of deformation. This frequency is related to the mean size of crystalline segments, which is about 20 mm. The segments are formed by clear fractures that go through the grain boundaries.

In this work, I suggest the following problem to be approached:

- To find a method of STF inversion from observed AE parameters taking into account a complex spectral structure of AE signals;

- To find a robust numerical criterion for the wave pattern recognition classified as a certain focal mechanism.

2. Normal modes of AE instrument

Let us consider a general problem of eigenfrequencies of a solid with free faces. In our case, we deal with lead-zirconate-titanate piezoceramic disk, which is a sensor of AE waves. It is encapsulated in a special instrument housing and transforms alternating deformation into an electrical signal. In order to estimate eigenfrequencies, which are the frequencies of normal modes, let us consider the fundamental Rayleigh-Lamb dispersion equation for symmetric and antisymmetric Lamb waves in the non-zero and zero cut-off frequencies [18, 19]. In the nonzero cut-off frequencies f_n the running plate waves turn into the standing dilatation D waves or shear S waves through the thickness h of plate (wave indexes are adopted from [18]):

$$f_n = \frac{2n-1}{2h} c_D \tag{1}$$

and

$$f_n = \frac{n}{h} c_S \tag{2}$$

for symmetric mode, and

$$f_n = \frac{n}{h} c_D \tag{3}$$

and

$$f_n = \frac{2n-1}{2h} c_S \tag{4}$$

for antisymmetric mode, where c_D and c_S are dilatation and shear wave velocities, respectively, n is an order of wave.

The second limiting condition relates to the lowest modes in plate [20, 21], which exist as propagating at any frequency no matter how low it is (the cut-off frequencies are zero). In this case, we are looking for the frequencies of standing waves arising along the diameter d dimension of a disk—they are the fundamental longitudinal and bending waves. Consideration of Rayleigh-Lamb dispersion equation for symmetric waves upon conditions that the wavelength λ is significantly greater than the plate thickness h and that the fundamental frequency f_d of standing longitudinal wave is determined by the wavelength $\lambda = 2d$, leads us to the equation of longitudinal wave frequency along the diameter d:

$$f_d = \frac{c_S\sqrt{c_D^2 - c_S^2}}{c_D d} \tag{5}$$

Applying the same conditions to the dispersion equation for antisymmetric wave, we obtain the fundamental frequency f_d of bending wave along d:

$$f_d = \frac{\pi}{\sqrt{3}} \frac{h}{2d^2} \frac{c_S}{c_D} \sqrt{c_D^2 - c_S^2} \tag{6}$$

where d is a diameter of piezoceramic disk.

In order to visualize free oscillations, the numerical simulation has been carried out by means of ANSYS mechanical finite element analysis software. The models of three thickness h modes of vibrations induced in the glass rectangular prisms of 1/5 aspect ratio are shown in **Figure 1** (left). I simplify the oscillation motions by picturing the displacement and stress distributions schematically for them (**Figure 1** (right)). The modes are the following: the 1st-order symmetric mode formed by the standing dilatation D wave (**Figure 1a**), corresponding frequency is expressed by Eq. (1); the 1st-order antisymmetric mode formed by the standing shear S wave (**Figure 1b**), corresponding frequency is expressed by Eq. (4); the 1st-order antisymmetric mode formed by the standing dilatation D wave (**Figure 1c**), corresponding frequency is expressed by Eq. (3). The latter is described by the full wave (see **Figure 1c**, right) in contrast to the former, where in one-half wavelength $h/2$ the displacements produce dilatation stresses, and in another—compression stresses.

AE waveforms apparently comprise the normal mode motions of the instrument, which are the matter of initial source transformation into a secondary acoustical process. In **Figure 2**, an example of high-frequency AE signal and its spectrum is performed. This example is extracted from AE data acquired during the deformation process of red granite from Blauenthal, Germany [22]. To study the nucleation process of brittle deformation, we conducted laboratory experiments with loading MTS frame on rock cores in GeoForschungsZentrum, Potsdam. To acquire AE data, the KRENZ PSO9070 transient recorder [23–25] has been used. This system digitized each AE waveform with a sampling rate of 5 MHz. The example in **Figure 2a** depicts a typical AE waveform occurring during the stage of brittle micro-fracture formation at the beginning of irreversible deformation. One can see that the informative frequencies range up to 2000 kHz (**Figure 2b**). Piezoceramic disk (Marco GmbH, Hermsdorf) of 1 mm thickness h and 5 mm diameter d has been used as a sensor of AE transducer. Let us estimate the normal mode frequencies of this sensor with the help of Eqs. (1)–(6), and see how they appear on the spectrum. Taking into account the wave velocities for that particular type of piezoceramics ($c_D \approx 4420$ m/s, $c_S \approx 3125$ m/s), we write the normal mode frequencies estimated by the Eqs. (1)–(6) in the order of increasing: zero-order antisymmetric mode along d (bending) 80 kHz; zero-order symmetric mode along d (longitudinal) 442 kHz; the standing shear S wave of the 1st-order

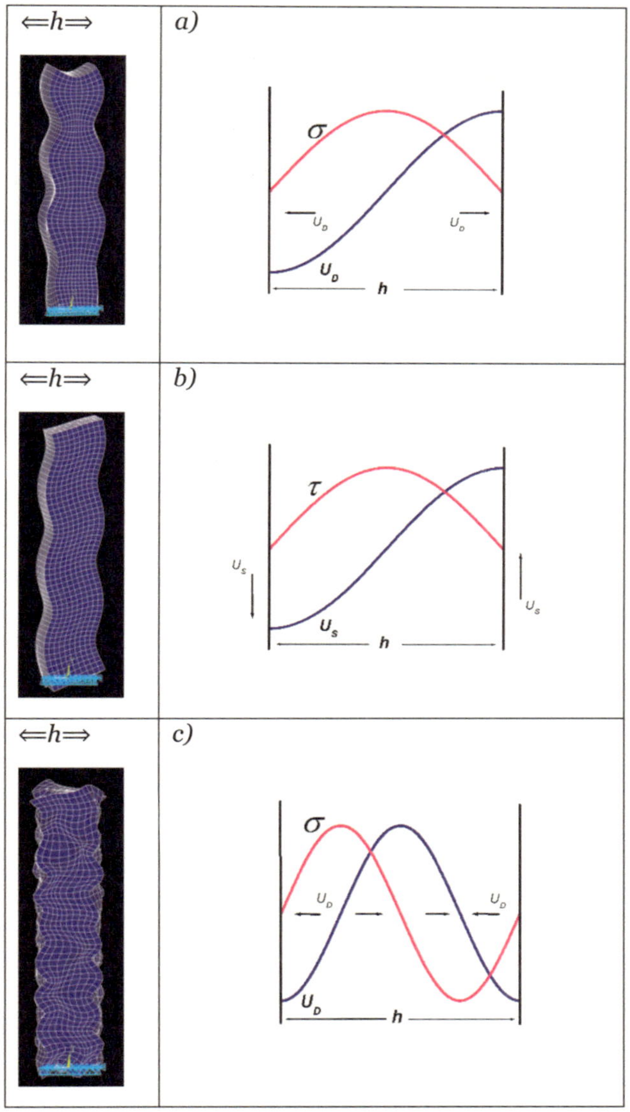

Figure 1.
Oscillation motions in standing waves through the lateral dimension h *(left), and the schematic view of displacements (blue line) and stresses (red line) in waves (right): a—the standing dilatation* D *wave of the 1st-order symmetric mode; b—the standing shear* S *wave of the 1st-order antisymmetric mode; c—the standing dilatation* D *wave of the 1st-order antisymmetric mode. σ—dilatation-compression stresses, τ—shear stresses,* U_D*—longitudinal displacements in wave, and* U_S*—transverse displacements in wave.*

antisymmetric mode (thickness h mode) 1563 kHz; the standing dilatation D wave of the 1st-order symmetric mode (thickness h mode) 2210 kHz; the standing shear S wave of the 1st-order symmetric mode (thickness h mode) 3125 kHz; and the standing dilatation D wave of the 1st-order antisymmetric mode (thickness h mode) 4420 kHz. I identify the dominant frequency on the spectrum in **Figure 2b** with longitudinal mode because it is close to the estimated value of 442 kHz. The reduced spectral value compared to the estimated one can result from exponential attenuation.

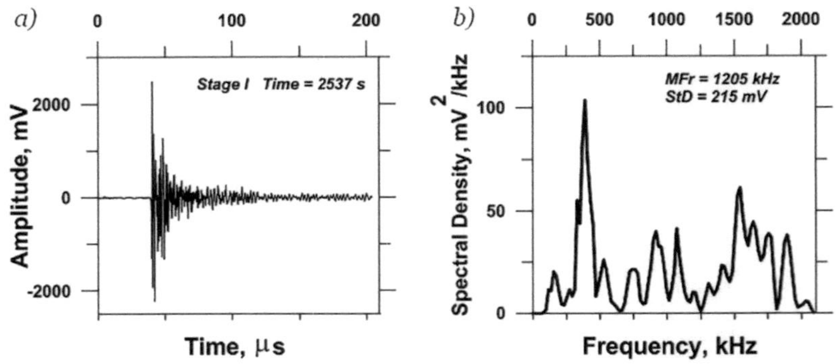

Figure 2.
Example of AE waveform, (a) and its spectrum and (b) selected from AE data of the red granite experiment (taken from [22]).

The next distinct peak on the spectrum (**Figure 2b**) appears in the vicinity of 1500 kHz and relates to the standing shear *S* wave of the 1st-order antisymmetric mode estimated as 1563 kHz. The lowest slight peak occurs in vicinity of 100 kHz and can be attributed to the bending mode characterized by estimated 80 kHz.

Thus, we can conclude that AE spectra contain two zero-order modes associated with diameter *d* of the sensor disk, and one evident mode associated with thickness *h*– the 1st-order antisymmetric mode formed by the standing shear *S* wave. Displacement and stress distributions of the latter are shown in **Figure 1b**. The energy of a wide-band AE source radiation partially transformed into a set of narrow-band modes.

Correspondence of estimated mode frequencies of piezo-disk to AE transducer eigenfrequencies we prove when analyzing the amplitude- and phase-frequency response characteristics performed in **Figure 3**. Calibration data of AE transducers was provided by the research team from GeoForschungsZentrum, Potsdam, in 2007. The scientists measured the response characteristic of each transducer with the network analyzer Advantest R3754A and performed in the form of decibel-frequency characteristics. Measurements are based on synthesizing of frequency sources, the frequency of which is swept to rapidly obtain amplitude and phase information of the receiver over a frequency band of interest. In these calibrations, a frequency band of 100–5000 kHz has been applied.

Looking at transducer characteristics (**Figure 3**), we find that close to the frequency 442 kHz of longitudinal *d* mode there is an explicit peak (red curve). At the same time, we observe the corresponding sharp drop of phase to the negative values (blue curve) which is an evidence of a strong resonance at this frequency. The group of peaks within the frequency range of 2000–3000 kHz can be presumably related to the 1st-order symmetric mode associated with thickness *h*. It is interesting to note that in the vicinity of 4420 kHz, which is a frequency of the standing dilatation *D* wave of the 1st-order antisymmetric mode, we observe a second drop of phase. This behavior of phase also reflects a strong resonance at this frequency. Oscillations for this certain mode are shown in **Figure 1c**.

In summary, the solutions of Rayleigh-Lamb equation in cut-off frequencies for an order $n \leq 1$ bring the modes explicitly detectable after impulse excitation. The normal mode of a linear system is a specific space distribution of displacements that cannot be decomposed into elementary distributions. Each normal mode is energy isolated from the other. Thus, an energy transmitted to the system by an impact can be estimated as a sum of energies of each normal mode. In [26, 27] it is shown that when radiated

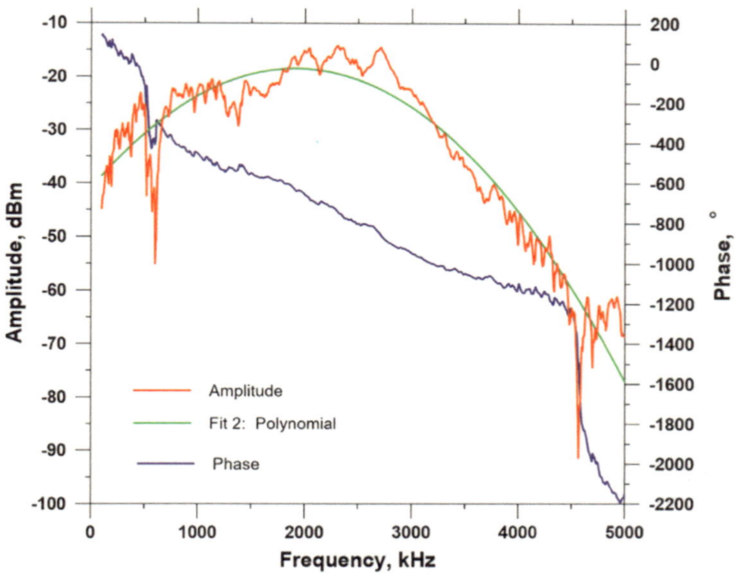

Figure 3.
Amplitude- and phase-frequency response of AE transducer type (GFZ, Potsdam): amplitude-frequency—red line, phase-frequency—blue line.

impulse duration is equal to a period of sample natural vibrations the total transmitted energy is converted into normal mode energy.

The forced response of a complex linear system «medium and seismic sensor» to the wide-band AE radiation is a superposition of normal modes. The intensity of each mode of excitation characterizes the STF of AE radiation.

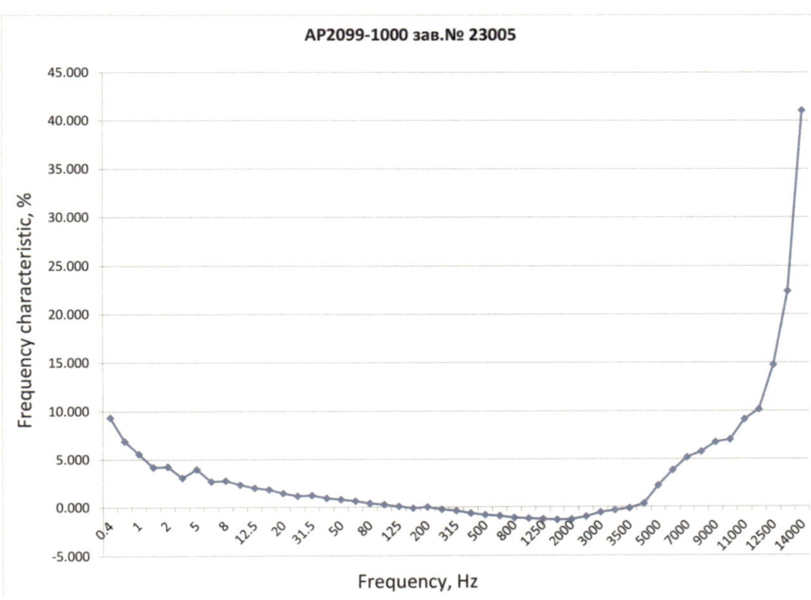

Figure 4.
Frequency characteristic of AE acceleration instrument (GlobalTest, Russia) for mine measurements.

At the end of this section, I introduce the response frequency characteristic of AE transducer type we are using in mine measurements. We will need it in Section 4. This is an acceleration instrument AP2099–1000 manufactured by the company GlobalTest, Russia, with a piezoelectric sensor embedded into the instrument. Frequency characteristic is presented in **Figure 4**. We see a strong increase in sensitivity on a frequency of 14 kHz, up to 40% of sensitivity on the reference frequency of 200 Hz (according to the calibration data provided by GlobalTest). Definitely, it is a result of transducer eigenfrequency excitation.

3. Spectral density and coherence function calculations

AE signal is a transient process that exists for a finite time T. The wave packet of such a process contains a sum of attenuating harmonic oscillations, the meaning of which is the normal modes of a system «medium and seismic sensor». Then it is reasonable to apply the Finite Fourier Transform and find Fourier coefficients for the function of T period [7, 8]. In executing this procedure one should take into account an additional scaling factor.

Calculations of spectral density and coherence function I perform by self-developed computer application. At first step, the basic Fast Fourier procedure produces on its output the real Re and imaginary Im components of discrete time series of Fourier transform pair:

$$\left.\begin{aligned} X_k &= \sum_{n=1}^{N} x_n \exp\left(\left(-i\frac{2\pi}{N}\right)(n-1)(k-1)\right) \\ x_n &= \sum_{k=1}^{N} X_k \exp\left(\left(i\frac{2\pi}{N}\right)(n-1)(k-1)\right) \end{aligned}\right\} \tag{7}$$

where X_k: direct Fourier transform; x_n: inverse Fourier transform; $n = 1, \dots, N$; $k = 1, \dots, N$; n: time counter; k: frequency counter; N: the number of samples of transient process. The correspondence of the real ReX_k and imaginary ImX_k components to Fourier coefficients a_k и b_k is expressed as follows:

$$\left.\begin{aligned} \mathrm{Re}\,X_k &= \frac{T}{2\Delta t}a_k = \frac{N}{2}a_k = \sum_{n=1}^{N} x_n \cos\left(\frac{2\pi}{N}(n-1)(k-1)\right) \\ \mathrm{Im}X_k &= \frac{T}{2\Delta t}b_k = \frac{N}{2}b_k = \sum_{n=1}^{N} x_n \sin\left(\frac{2\pi}{N}(n-1)(k-1)\right) \end{aligned}\right\} \tag{8}$$

where Δt: sampling period; $T = N\Delta t$: signal length. X_k is a Finite Fourier Transform on discrete frequencies $f_k = k/T$.

Further, it is necessary to obtain physically meaningful estimations of spectral densities. For discrete x_n transient process with the sampling period Δt the initial spectral density evaluation we derive as follows:

$$\tilde{G}_x(f_k) = \frac{2}{\Delta t N}\left(\Delta t^2(\mathrm{Re}\,X_k)^2 + \Delta t^2(\mathrm{Im}X_k)^2\right) = \frac{2\Delta t}{N}|X_k|^2 \tag{9}$$

where $|X_k|^2 = (\mathrm{Re}\ X_k)^2 + (\mathrm{Im}\ X_k)^2$: squared modulus of Finite Fourier Transform.

The scaling factor $2\Delta t/N$ introduces the Finite Fourier Transform as a weight of a frequency component f_k normalized to the full-scale range. Thus, the dimension of spectral density becomes «Amplitude²/Hz».

With the help of spectral density, it becomes suitable to make an estimate of transient energy in any frequency band defined by certain measurement conditions. For this purpose, we invoke the Parseval theorem [28], which poses the relation between the mean square of any transient process x_n and the spectral integral:

$$\frac{1}{N}\sum_{n=1}^{N} x_n^2 = \frac{1}{2}\sum_{k=1}^{N/2}\left(a_k^2 + b_k^2\right) \tag{10}$$

The computer application also calculates the frequency-dependent correlation of the signals x_n and y_n arising in the stream of AE events by means of the coherence function. Radiation sources are coherent if they radiate harmonic waves of time-constant phase shift, and their sum gives a wave of the same frequency. For that, we should satisfy a requirement for the phases of this certain frequency to be non-random when dividing each of the signals into n_d equal segments. Thus, to carry out coherence analysis, it is necessary to derive statistically significant values of spectral densities. For the discrete time series x_n we obtain:

$$G_{XX}(f_k) = \frac{2}{n_d T_j}\sum_{j=1}^{n_d}\left(\Delta t^2 (\mathrm{Re}\,X_{kj})^2 + \Delta t^2 (\mathrm{Im}X_{kj})^2\right) = \frac{2\Delta t}{n_d N_j}\sum_{j=1}^{n_d}\left|X_{kj}\right|^2 \tag{11}$$

where $j = 1, \ldots, n_d$; $T_j = \Delta t N_j$: segment length. The same we obtain for the second time series y_n as a value $G_{YY}(f_k)$.

Further, we should calculate the cross-spectral density $G_{XY}(f_k)$, which is a complex value unlike the values of $G_{XX}(f_k)$ and $G_{YY}(f_k)$. The program implements these calculations in the following way. Firstly, the real $\mathrm{Re}XY_{kj}$ and imaginary $\mathrm{Im}XY_{kj}$ components of cross-spectrum Finite Fourier Transform are derived:

$$\left.\begin{aligned} \mathrm{Re}\,XY_{kj} &= \mathrm{Re}\,X_{kj}\cdot \mathrm{Re}\,Y_{kj} + \mathrm{Im}X_{kj}\cdot \mathrm{Im}Y_{kj} \\ \mathrm{Im}XY_{kj} &= \mathrm{Re}\,X_{kj}\cdot \mathrm{Im}Y_{kj} - \mathrm{Im}X_{kj}\cdot \mathrm{Re}\,Y_{kj} \end{aligned}\right\} \tag{12}$$

where X_{kj} and Y_{kj} are the Finite Fourier Transforms on discrete frequencies $f_k = k/T_j$ of the time series x_n and y_n for a segment j. Afterward, we find cross-spectrum modulus:

$$\left|G_{XY}(f_k)\right| = \frac{2\Delta t}{n_d N_j}\sqrt{\left(\sum_{j=1}^{n_d}\mathrm{Re}\,XY_{kj}\right)^2 + \left(\sum_{j=1}^{n_d}\mathrm{Im}XY_{kj}\right)^2} \tag{13}$$

Finally, we obtain the coherence function $\gamma^2_{XY}(f_k)$ on discrete frequencies f_k. To explicitly specify the coherence function, I suggest the following notations:

$A = \mathrm{Re}X_{kj}$; $B = \mathrm{Im}X_{kj}$; $C = \mathrm{Re}Y_{kj}$; $D = \mathrm{Im}Y_{kj}$

Then, it gives expression in the following form:

$$\gamma^2_{XY}(f_k) = \frac{\left(\sum(AC+BD)\right)^2 + \left(\sum(AD-BC)\right)^2}{\left(\sum(A^2+B^2)\right)\left(\sum(C^2+D^2)\right)} \tag{14}$$

Here, we mean a summation over a number n_d of segments on each frequency f_k. In such representation of the coherence function, we can assure that, in case two signals are coherent on a certain frequency f_k, that is $A = C$ and $B = D$, the coherence function gets its maximum value at this frequency equal to 1.

4. Examples of stick-slip seismicity control in the mines of the unique Khibiny appatite deposit, Kola peninsula, Russia

Rock massif of Khibiny appatite deposit is characterized by a complex geological structure—this is a system of joints running in different directions. As a result of geological processes, a block-like structure was formed, furthermore, the faults are scale ranked as large-block (extension is hundreds of meters) and small-block (extension less than tens of meters). In terms of geological specifics, the faults can be filled with secondary penetration minerals (e.g., monchikite dike), or empty. In accordance with these geological features, we introduce the following terms of fault surface contact—the contact «rock-dyke» and «rock-rock». This we need for spectrum classification of *stick-slip* process.

To describe the conditions of mining excavation a few words should be mentioned. "Apatit" is a company that developed this unique apatite-nepheline ore deposit in the mountain region since 1929. It produces a high-grade phosphate rock. At the same time, conditions of mining become complicated because of high tectonic stress resulting in different kinds of dynamic rock activity, such as microcrack nucleation and *stick-slip* events, which in turn lead to failure incident. In general, this deposit is classified as a rock-burst hazardous one.

In 2015, during AE monitor procedure in one of the underground mining tunnels, we registered a sequence of specific events in the course of a week [29]. The main feature of the rock massif on site was a monchikite dike on the contact with appatite-nepheline ore. It should be noted that the same signature of AE signals has been revealed at the post-strength stage during the laboratory test of the red granite core [22] (we mentioned in Section 2). This stage is characterized as a frictional sliding or shearing on the fault surfaces. I show an example of such AE doublet in **Figure 5a** and **b**, as well as their cross-spectrum (black line), and coherence function (blue line), c).

The spectral structure of highlighted AE signals is characterized by two well-defined dominant frequencies *DFr*—117 and 352 kHz (**Figure 5b**). The coherence function indicates high correlation values (close to 1) within the frequency range of cross-spectrum significant values—100–500 kHz (**Figure 5c**). It is interesting to point out that the upper limit of this range coincides with the frequency of zero-order symmetric mode 442 kHz of the piezoelectric sensor (see Section 2). Thus, we may conclude that the signals of this particular type are of similar radiation nature, which is qualified by two independent spectral components and П-pattern coherence function. According to the spectral pattern similarity of these laboratory AE signals to the seismicity registered in mine in the vicinity of the contact «rock-dyke» [29] I have designated the mining AE events to be attributed to *stick-slip* process. Tectonic activity in this region can generate frictional sliding on the contact «rock-dyke» the same as in laboratory rock cores pressure initiates the shearing on fault surfaces at the post-strength stage.

After this finding, we proceeded to monitor AE process in different places of mine close to dike occurrence. The result was the same selected type of signal. In this work, I perform the results of AE monitoring carried out by the special mining devices of

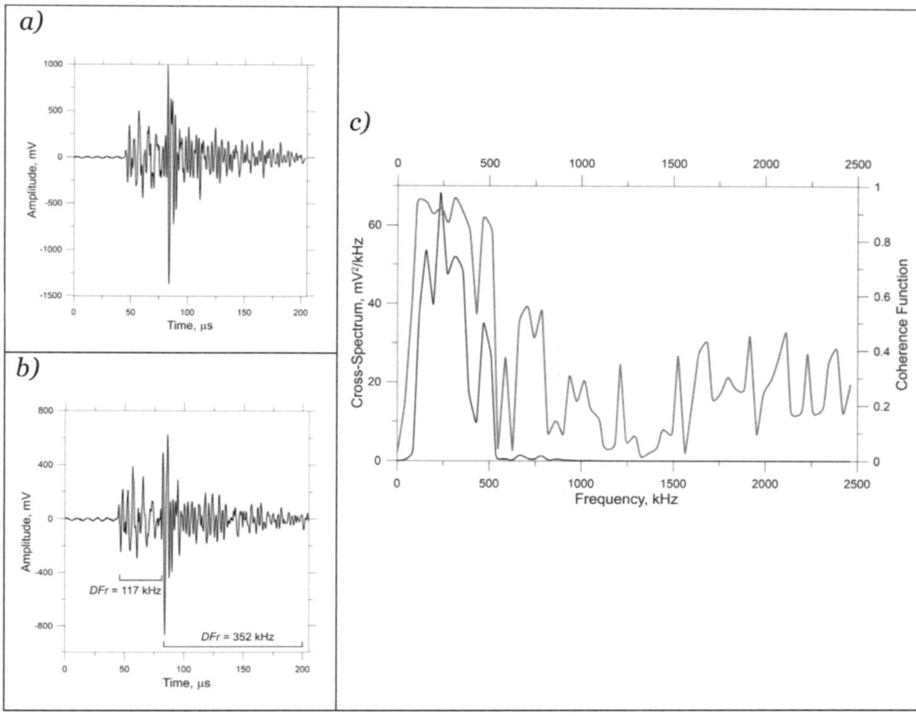

Figure 5.
AE doublet acquired during the laboratory study of the deformation process of red granite core [22] on the post-strength stage of relative shearing on fault surfaces, (a) and (b); cross-spectrum (black line) and coherence function (blue line) derived for these signals, (c).

Prognoz-L product line (**Figure 6a**). These devices are developing in the mining institute of the far-eastern division of the Russian Academy of Science [30]. The acceleration instrument AP2099–1000, GlobalTest, is used as AE transducer (see Section 2). The measurement frequency range is 0.5–30 kHz, the eigenfrequency is equal to 14 kHz (**Figure 4**). The nominal value of the electroacoustic transfer factor on the reference frequency is 100 $mV/m/s^2$.

Let us consider the measurement series conducted in the period of 22.10.2021–12.11.2021 aimed to control the stability of rock massif after a strong seismic event, which was recorded by the seismic acquisition system of mine. The rocks of the massif are strongly fractured, and the monchikite dike of 0.6 m thickness occurs with an inclined angle of 60°–65° (**Figure 6b**).

Figures 7 and **8** perform the capture samples of AE process and their spectrum moduli. It is evident that both spectra consist of two independent spectral components: low-frequency narrow-band component corresponding to a quasi-stationary monochromatic process and high-frequency wide-band component described by the main lobe and two side lobes. According to the spectral theory [31], the latter (underlined by a red transparent line) is produced by a sine fragment of frequency which corresponds to the main lobe peak. Moreover, the spectral half-width of the main lobe is 2 kHz and it characterizes approximate value of the envelope length (about 0.5 μs). It means that this process has an apparent impulse (non-periodic) property with a basic frequency of 14 kHz related to the transducer eigenfrequency

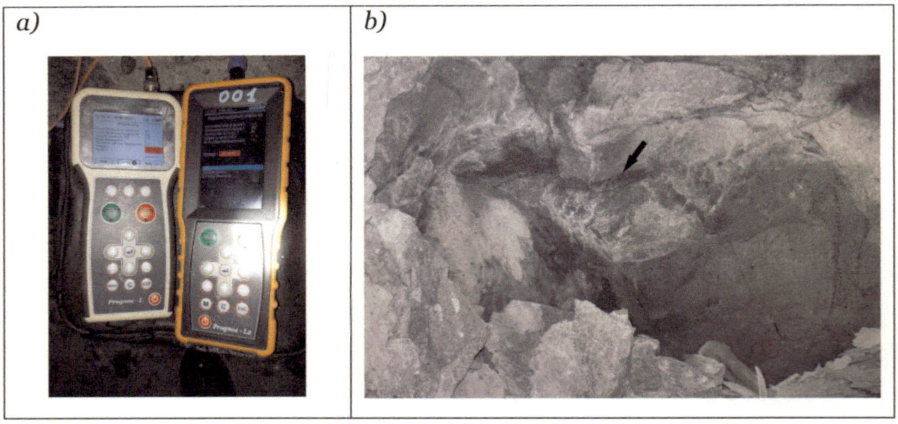

Figure 6.
Mining devices Prognoz-L for seismicity control, (a); a photograph of rock massif, view from the tunnel, (b) the tunnel exhibits a great number of fractures and a trace of monchikite dyke, marked by the black arrow.

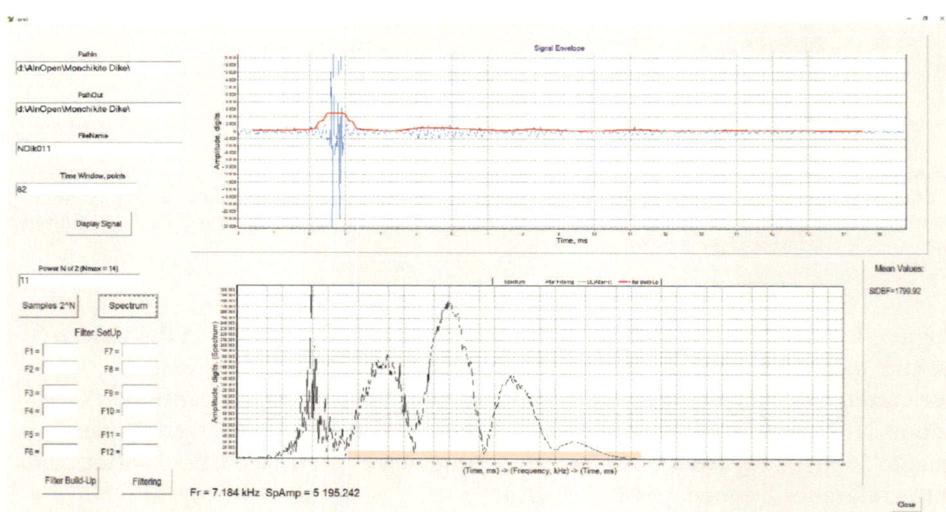

Figure 7.
Two-component spectrum of a typical AE stick-slip *signal presumably generated on the contact «rock-dyke»: signal—blue line; trapezium-like envelope—red line (upper part of the interface); spectrum—black line (lower part of the interface): the frequency range associated with the envelope and inserted frequency 14 kHz is underlined by red transparent line—a fast radiation; the mode of 4.89 kHz—a slow radiation.*

(**Figure 4**). In other words, the normal mode effect shifts the envelope function of *sinc* (ω) in the frequency ω domain by the value of normal mode.

Thus, **Figures 7** and **8** illustrate the same property of wave source. The only difference we can emphasize is that the two-component spectrum pattern in **Figure 8** is shifted to the higher frequencies in comparison with the pattern in **Figure 7**. The frequency of monochromatic component moves from 4.89 kHz position to 7.33 kHz position, and the onset of the impulse spectrum moves from 7.184 to 9.862 kHz. Interpretation of two-component spectrum feature is based on the models of

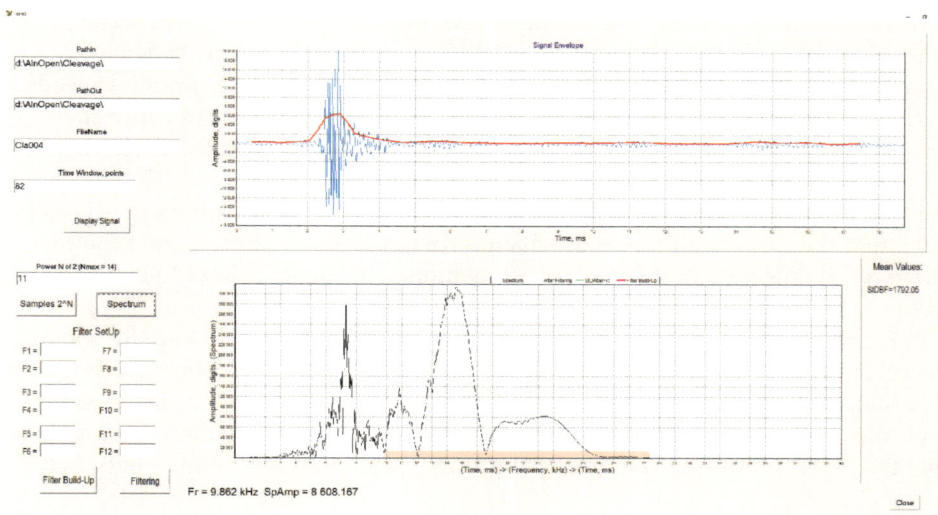

Figure 8.
Two-component spectrum of a typical AE stick-slip *signal presumably generated on the contact «rock-rock»: signal—blue line; trapezium-like envelope—red line (upper part of the interface); spectrum—black line (lower part of the interface): the frequency range associated with the envelope and inserted frequency 14.59 kHz is underlined by red transparent line—a fast radiation; the mode of 7.33 kHz—a slow radiation.*

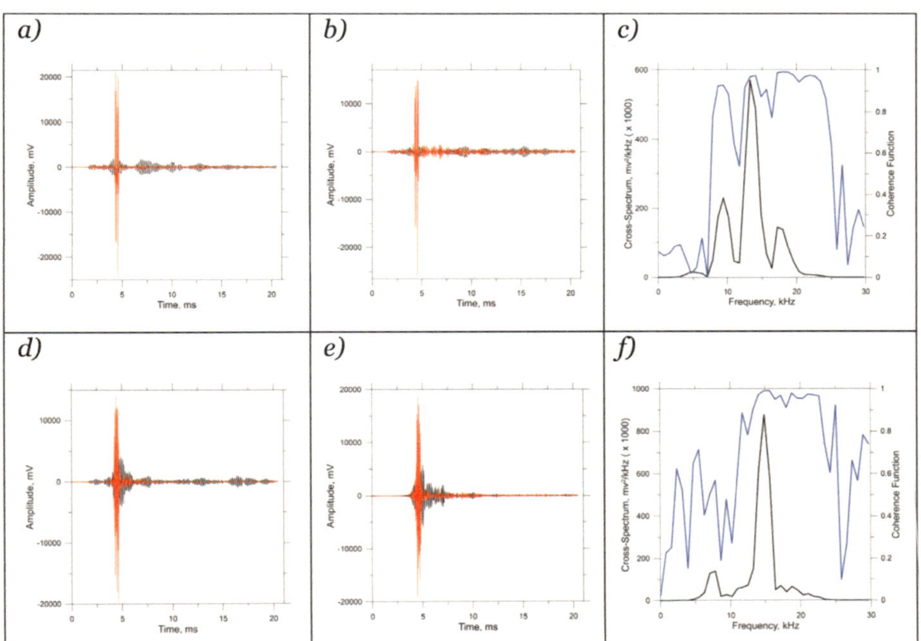

Figure 9.
AE doublet of impulse (non-periodic) signals of stick-slip *on the contact «rock-dyke», (a) and (b); their cross-spectrum (black line) and coherence function (blue line), (c); AE doublet of impulse (non-periodic) signals of* stick-slip *on the contact «rock-rock», (d) and (e); their cross-spectrum (black line) and coherence function (blue line), (f).*

complex sources [1] one of which is the model of asperities on the fault surfaces. In the focal region of a *stick-slip* source, two different processes occur. When a force is applied to one of the rock blocks, its elastic strain energy increases until it exceeds the strength of asperities on the contact «rock-dyke» or «rock-rock». Once the strength is achieved and asperities are fractured, the tangential stress drop occurs on the fault surfaces resulting in the instantaneous slip of a block, which is accompanied by elastic wave radiation. Fast radiation (short pulse) corresponds to asperity fracturing, the spectrum pattern of which is designated by a shifted *sinc*-function. Whereas the slow radiation (low-frequency monochromatic) relates to quasi-stationary frictional sliding.

In **Figure 9**, I show the examples of AE *stick-slip* signals initiated on the contact «rock-dyke», (a) and (b), their cross-spectrum (black line) and coherence function (blue line), (c), and the examples of AE *stick-slip* signals initiated on the contact «rock-rock», (d) and (e), their cross-spectrum (black line) and coherence function (blue line), (f). Moreover, the signals are decomposed into two components: black

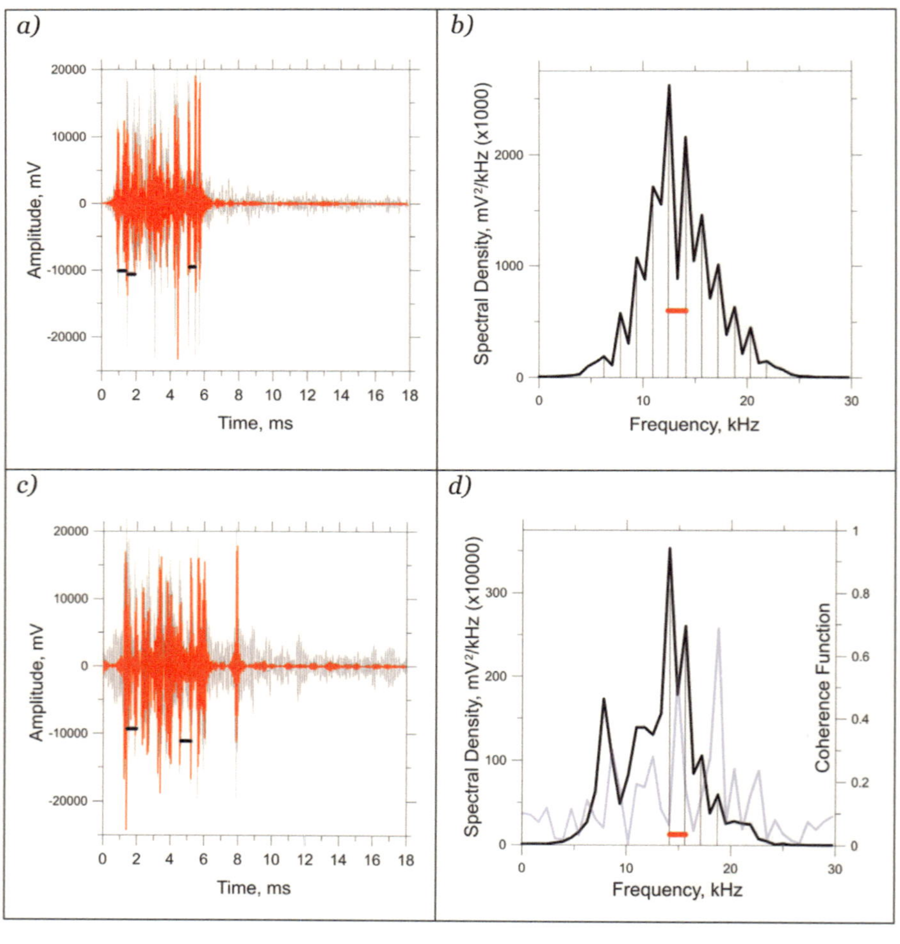

Figure 10.
AE stick-slip *signals associated with multiple asperity failure on the contact «rock-rock» recorded in rock massif of small-block faults, (a) and (c); their spectral densities presented by discrete spectra, (b) and (d); coherence function for these signals (blue line), (d).*

color depicts the narrow-band process related to quasi-stationary frictional sliding and red line depicts impulse wide-band process related to asperity fracturing on fault surfaces. It appears that the coherence function is described by Π-pattern, the same as observed during the red granite deformation process (**Figure 5**). It is worth noting that one can see the shift of the lower limit of Π-pattern toward higher frequencies for the *stick-slip* process on the contact of «rock-rock» (**Figure 9f**, blue line) in comparison with «rock-dyke» (**Figure 9c**, blue line). This fact reflects the shifting of the left side lobe lower limit for «rock-rock» *stick-slip* (**Figure 8**) relative to «rock-dyke» *stick-slip* (**Figure 7**).

The evidence of asperity influence on AE process during *stick-slip* is experimentally shown, for example, in Ref. [32]. There the authors observe several bursts of AE activity against the background of a gradual AE increase prior to the main slip event. These AE bursts explained by the failure of grain-scale asperities on rough surfaces of a fault.

Let us proceed to the quasi-periodic radiation of AE *stick-slip* source, which is characterized by multiple asperity failure. Evidently, in this case, the coherence function Π-pattern will be destroyed (**Figure 10d**, blue line) because the phases of harmonics characterizing the impulse radiation become random resulting in the reduction of the values of coherence function. At the same time, the calculations of statistically significant values of spectral densities reveal the transformation of the continuous spectrum (**Figures 7** and **8**) into a discrete one (**Figure 10b** and **d**) in the area of transducer eigenfrequency 14 kHz. The spectrum discrete nature is defined by frequency increment Δf, which is an inverse value of a period T which in turn characterizes a sequence of pulses generated by asperity fracturing (**Figure 10a** and **c**, the filtered impulse component is depicted by red color). Based on spectral theory [31], this transformation can be illustrated as it is shown in **Figure 11**. The spectrum (**Figure 11b**) of the synthetic signal is calculated, which is composed of attenuating bipolar pulses with a period 10 μs and length 2.5 μs (**Figure 11a**). As a result, the continuous spectrum of a single pulse (**Figure 11b**,

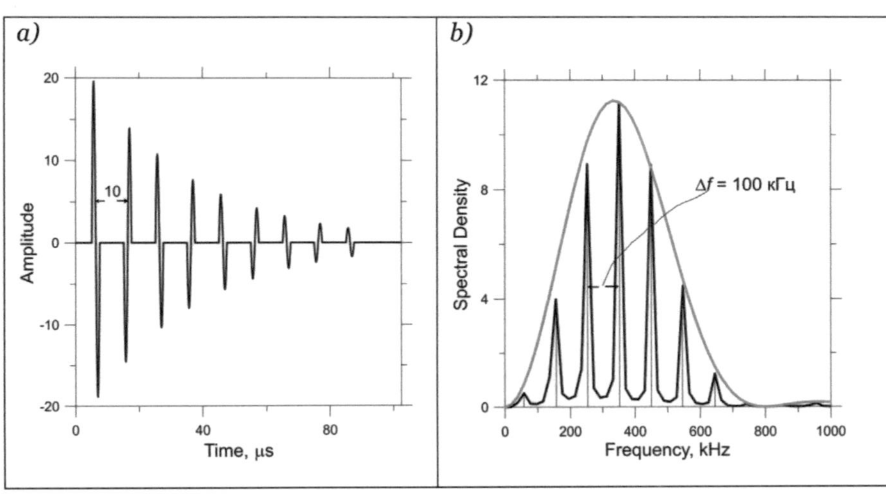

Figure 11.
Illustration of continuous spectrum transformation into discrete spectrum due to pulse periodic repeat: (a) synthetic signal composed of attenuating bipolar pulses with period 10 μs and length 2.5 μs; (b) discrete spectrum of pulse periodic repeat (black line), and continuous spectrum of the single pulse (gray line).

gray line) turns into a discrete spectrum of periodic sequence of pulses (**Figure 11b**, black line).

In the end of this section, it should be noted that AE *stick-slip* signal classification according to the contact «rock-dyke» or «rock-rock» is based on two items: geological characteristic of rock massif on site, and spectral displacement to higher frequencies talking about the contact «rock-rock». I argue the latter by the fact that the higher the frequencies are, the shorter is STF, which in turn results from *stick-slip* radiation in small-block faults (on the contact «rock-rock»).

5. Conclusions

Based on spectral-coherency analysis of AE signals on laboratory scale and mining scale, we can conclude the following:

- AE sensor normal modes strongly influence a radiated source wave modifying it into a set of narrow-band processes;
- Thus, we may convolve an origin AE wave $A(f)$ with normal modes $NM(f)$ of a system «medium and seismic sensor» and derive a resultant output signal: $S(f) = A(f) \cdot NM(f)$;
- П-pattern of coherence function has been revealed, which presumably specifies AE doublet of nonperiodic signals of *stick-slip* type, which can be easily selected in AE data;
- Periodic behavior in *stick-slip* source is detected by discrete spectrum obtained by calculations of statistically significant values of spectral densities, which in turn is evidence of multiple asperity failure on fault surfaces.

The suggested approach of spectral-coherency analysis of AE control in mines can be applied to the development of new algorithms for critical-stage automatic detection.

Acknowledgements

I thank Prof. Dr. Georg Dresen from GeoForschungsZentrum, Potsdam (Germany) for providing me with the calibration data of AE transducers.

I appreciate the help and support of the team of Rock Hazard Prevention Service of JSC «Apatit» (Kola Peninsula, Russia) in conducting of AE measurements under mining conditions and interpretation.

I thank engineers from Mining Institute of the Russian Academy of Sciences (Far East, Russia) for consulting me on the technical issues of portable geoacoustic device «Prognoz-L» operation.

Conflict of interest

The author declares no conflict of interest.

Author details

Alexander Rozanov
Saint Petersburg Mining University, Saint Petersburg, Russia

*Address all correspondence to: aleksander.rozanov@mail.ru

References

[1] Udias A. Principles of Seismology. UK: Cambridge University Press; 1999. 468 p. DOI: 10.1016/S0012-8252(00)00040-4

[2] Frenkel YI. Introduction to the Theory of Metals. 2nd ed. Moscow: State Publishing House of Technic-Academic Literature; 1950. 353 p. [in Russian]

[3] Hertzberg R, W, Vinci RP, Hertzberg JL. Deformation and Fracture Mechanics of Engineering Materials. 5th ed. New York: John Wiley & Sons, Inc.; 2013. 747 p. DOI: 10.1115/1.3443416

[4] Reches Z, Lockner D. Nucleation and growth of faults in brittle rocks. Journal of Geophysical Research. 1994; **99**(B9):18159-18173. DOI: 10.1029/94JB00115

[5] Lockner D. Rock Failure. US: The American Geophysical Union; 1995

[6] Bolton DC, Shreedharan S, Rivière J, Marone C. Acoustic energy release during the laboratory seismic cycle: Insights on laboratory earthquake precursors and prediction. Journal of Geophysical Research: Solid Earth. 2020; **125**:1-17. DOI: 10.1029/2019JB018975

[7] Bendat JS, Piersol AG. Engineering Applications of Correlation and Spectral Analysis. New York: John Wiley and Sons; 1980

[8] Bendat JS, Piersol AG. Random Data: Analysis and Measurement Procedures. New York: Wiley-Interscience; 1974. DOI: 10.1002/9781118032428

[9] Boler FM, Spetzler H. Radiated seismic energy and strain energy release in laboratory dynamic tensile fracture. PAGEOPH. 1986;**124**:759-772. DOI: 10.1007/BF00879609

[10] Stanchits SA, Savel'ev VN, Boler FM. Recording stresses and strains created by elastic waves. Izvestiya, Earth Physics. 1987;**23**(3):258-261

[11] Rozanov AO. Microseismic event spectrum control and strain energy release in stressed rocks. Search and Discovery Article. 2012;**#40897**:4

[12] Rozanov AO, Savel'ev VN, Stanchits SA. Spectrum analysis of pulse-induced acoustic field in jointed and faulted rock formations. In: Proceedings of the 57th Conference, European Association of Geoscientists and Engineers, Extended Abstracts. Netherlands: European Association of Geoscientists and Engineers; 1995. Paper B004. DOI: 10.3997/2214-4609.201409205

[13] Krauklis PV, Krauklis LA. Unexpected wave in layered media with loosely bonded interfaces. In: Proceedings of the 58th Conference, European Association of Geoscientists and Engineers, Extended Abstracts. Netherlands: European Association of Geoscientists and Engineers; 1996. Paper P075. DOI: 10.3997/2214-4609.201408837

[14] Bottelin P, Levy C, Baillet L, Jongmans D, Gueguen P. Modal and thermal analysis of les arches unstable rock column (Vercors massif, French Alps). Geophysical Journal International. 2013;**194**:849-858. DOI: 10.1093/gji/ggt046

[15] Dietze M, Krautblatter M, Illien L, Hovius N. Seismic constraints on rock damaging related to a failing mountain peak: The Hochvogel, Allgau. Earth Surface Processes and Landforms. 2021; **46**:417-429. DOI: 10.1002/esp.5034

[16] Rozanov A, Zang A, Wagner C, Dresen G. Acoustic frequency signatures of laboratory fractured rocks. In: Proceedings of the 63rd Conference, European Association of Geoscientists and Engineers, Extended Abstracts. Netherlands: European Association of Geoscientists and Engineers; 2001. Paper P036. DOI: 10.3997/2214-4609-pdb.15.P036

[17] Xian C, Nolte DD, Pyrak-Notle LJ. Compressional waves guided between parallel fractures. International Journal of Rock Mechanics & Mining Sciences. 2001;**38**:765-776

[18] Mason WP, editor. Physical Acoustics. Vol. 1. Methods and Devices. Part A. 1st ed. New York and London: Academic Press; 1964. 592 p

[19] Isakovich MA. General Acoustics [in Russian]. Общая акустика. Moscow: Nauka; 1973. 496 p

[20] Krasylnikov VA, Krylov VV. Introduction to Physical Acoustics, [in Russian]. Moscow: Nauka; 1984. 400 p. DOI: 10.13140/RG.2.1.1129.0086

[21] Brekhovskikh LM, Goncharov VV. Continuum Mechanics in Application to Wave Theory, [in Russian]. Moscow: Nauka; 1982. 325 p

[22] Rozanov AO. Spectral and correlation analysis of acoustic emission waveforms and failure micromechanics in rocks. In: Proc. Int. Geophys. Conf. & Exhibition on Geophysics of the 21st Century, Extended Abstracts. Netherlands: European Association of Geoscientists and Engineers; 2003. DOI: 10.3997/2214-4609-pdb.38.F071

[23] Zang A, Wagner FC, Dresen G. Acoustic emission, microstructure, and damage model of dry and wet sandstone stressed to failure. Journal of Geophysical Research. 1996;**101**: 17507-17521. DOI: 10.1029/96JB01189

[24] Zang A, Wagner FC, Stanchits S, Dresen G, Andresen R, Haidekker MA. Source analysis of acoustic emissions in Aue granite cores under symmetric and asymmetric compression loads. Geophysical Journal International. 1998; **135**:1113-1130. DOI: 10.1046/j.1365-246X.1998.00706.x

[25] Zang A, Wagner FC, Stanchits S, Janssen C, Dresen G. Fracture process zone in granite. Journal of Geophysical Research. 2000;**105**(B10): 23651-23661. DOI: 10.1029/2000JB900239

[26] Rozanov AO, Kuksenko VS, Savel'ev VN, Stanchits SA, Pikulin VA. Estimating the energy of a source of elastic waves from the acoustic signal parameters. Technical Physics Letters. 1993;**19**(2):113-114

[27] Rozanov AO, Kuksenko VS, Savel'ev VN, Stanchits SA. Elastic-energy spectra produced in a solid by a pulsed radiation source. Physics of the Solid State. 1994; **36**(11):1797-1802

[28] Korn GA. Mathematical Handbook for Scientists and Engineers. New York Toronto London: McGraw-Hill Book Company, Inc; 1961

[29] Rasskazov IY, Tsirel SV, Rozanov AO, Tereshkin AA, Gladyr AV. Application of acoustic measurement data to characterize initiation and development of disintegration focus in a rock mass. Journal of Mining Science. 2017;**53**(2):224-231. DOI: 10.1134/S1062739117022055

[30] Rasskazov IY, Migunov DS, Anikin PA, Gladyr AV, Tereshkin AA, Zhelnin DO. New-generation portable

geoacoustic instrument for rockburst hazard assessment. Journal of Mining Science. 2015;**51**(3):614-623. DOI: 10.1134/S1062739115030278

[31] Kharkevich AA. Spectra and Analysis. US: Springer; 2013. p. 222. ISBN: 1489948651, 9781489948656

[32] Dresen G, Kwiatek G, Goebel T, Ben-Zion Y. Seismic and aseismic preparatory processes before large stick-slip failure. Pure and Applied Geophysics. 2020;**177**:5741-5760. DOI: 10.1007/s00024-020-02605-x

Chapter 4

Low-Frequency Bandgaps by Topological Acoustic Black Holes

Jie Deng

Abstract

Nowadays, acoustic black holes (ABHs) are very popular for producing efficient vibration reduction at high frequencies in combination with some damping mechanisms. However, its low-frequency performance is hard to improve since the ABH effect principally occurs beyond its cut-on frequency. Fortunately, periodic ABH configuration offers some bandgaps below that frequency for wave attenuation. In this chapter, a topological ABH structure is suggested to produce a new bandgap at very low frequencies, by taking a supercell and decreasing the ABH distance. The wave and Rayleigh-Ritz method (WRRM) is adopted to compute the complex dispersion curves. Examinations of the dispersion curves and transmissibilities confirm the efficiency of the low-frequency vibration reduction capability of the proposed topological ABHs.

Keywords: acoustic black holes, topological bandgaps, low frequencies, vibration reduction, complex dispersion curves

1. Introduction

Acoustic black holes (ABHs) in mechanics are very efficient for wave manipulation in structures. By decreasing structural thickness following a power law, the impinging wave can be significantly retarded and concentrated as it propagates to the tip. Usually, in the vicinity of the ABH tip, some damping mechanisms [1, 2] are adopted; thus, the vibration energy is converted to, for example, heat, resulting in highly efficient vibration attenuation. However, there exists a cut-on frequency that relies on the ABH size. It is generally thought that the ABH effect can only be efficient above that frequency. While in practical engineering situations the tough task is related to low frequencies. Therefore, it is imperative to ameliorate the low-frequency performance of ABH structures.

Periodic distribution of ABHs [3–5], a new type of phononic crystal/metamaterial, could produce some bandgaps below the cut-on frequency, constituting a promising approach for low-frequency vibration reduction. This concept is first investigated in Refs. [6–8], where the purpose is to modulate the wavefront of bending waves in plates. Later, a periodic ABH beam is proposed [9], reporting that many locally resonant low-frequency bandgaps are formed below the cut-on frequency. Recently, it has been unveiled that these bandgaps are caused by Bragg scattering [10].

Interestingly, a compound ABH beam structure (also called double-leaf ABH beam) has been extensively studied [11–13], due to that many broad bandgaps can be found. In Ref. [14], very low bangaps are obtained by modifying the ABH parameters. Nevertheless, it is very tricky to obtain complete bandgaps in plates. In Ref. [15], a circular double-leaf ABH structure is proposed, showing that complete and sub-wavelength bandgaps are found. While in Ref. [16], a strip ABH is proposed, providing broad bandgaps and showing promising application for wave manipulation in plates. On the other hand, periodically placing resonators on the ABH plate has also been investigated [17, 18], where a very low bandgap is formed to suppress the first formant of the plate. Not only that, nonlinear effect has been applyed for ABHs, taking the advantage of energy transfer from low to high frequencies [19]. Recently, as an alternative of embedded ABHs, the additive ABHs have been widely explored, showing that the added mass could somehow facilitate low-frequency wave reduction [20–23].

To further improve the low-frequency performance of embedded ABHs, we must resort to the combination of different physics. Inspired by topological metamaterials, in the current chapter, we propose topological ABHs to generalize a low-frequency bandgap, by adjusting the ABH distance in a supercell, as illustrated in **Figure 1**. It is worthwhile mentioning that the concept of topology has been introduced in previous works [24, 25]. However, the purpose of those works is concentrated on the presence of robust edge or surface states within the bandgap. Very differently, in the current chapter, we exploit the low-frequency bandgap generalized by the topological ABHs, to further enhance the low-frequency performance of ABHs. Please note that it is easy to prove the topological property by showing, for example, interface states or topological symmetry. We will overlook that because the focus is only placed on generating low-frequency bandgaps.

To unveil the properties of the proposed topological ABHs, we adopt our previously established wave and Rayleigh-Ritz method (WRRM, see [18, 26]) to recover the complex dispersion curves, where the imaginary part stands for the wave attenuation (including damping and bandgaps). Furthermore, the transmissibilities of finite structures will also be tested, to prove the existence of low-frequency topological bandgap.

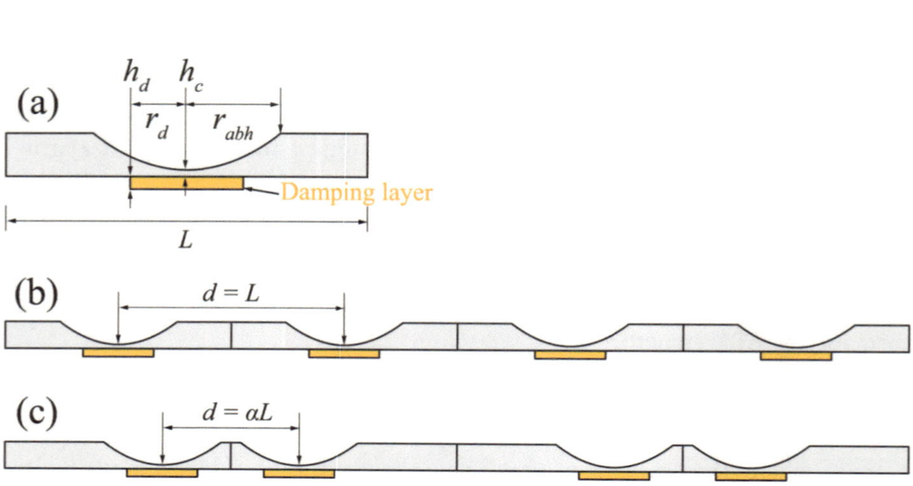

Figure 1.
Illustration of (a) a unit ABH cell with detailed geometry parameters, (b) periodic ABHs, and (c) topological ABHs.

2. Wave and Rayleigh-Ritz method

In this section, we will place the focus on the wave and Rayleigh-Ritz (WRRM) to compute the complex dispersion curves. For the ease of exposition, we take the Euler-Bernoulli assumption, such that the bending displacement of the beam can be expanded by a series of basis functions,

$$w(x,t) = \sum_{i=1}^{n} a_i(t)\varphi_i(x) := \boldsymbol{\varphi}^\top(x)\boldsymbol{a}(t), \tag{1}$$

where $\boldsymbol{a}$ is the unknown coefficient vector to be determined and $\boldsymbol{\varphi}$ signifies a Gaussian function vector (see the Gaussian expansion method, GEM [27, 28]).

One can continue in the Rayleigh-Ritz framework to reach the following equation of motion,

$$\left(\boldsymbol{K} - \omega^2\boldsymbol{M}\right)\boldsymbol{A} = \boldsymbol{0}, \tag{2}$$

where $\boldsymbol{M}$ and $\boldsymbol{K}$ respectively denote the mass and stiffness matrices. Please note that in Eq. (2) we have taken $\boldsymbol{a} = \boldsymbol{A}\exp(\mathrm{i}\omega t)$, where ω represents the angular frequency.

The following step is to impose the Bloch-Floquet periodic conditions at the cell boundaries,

$$w_1 = \lambda w_2, \tag{3}$$

$$\partial_x w_1 = \lambda \partial_x w_2, \tag{4}$$

where w_1 and $\partial_x w_1$ respectively represent the displacement and rotational angle at the left end, and w_2 and $\partial_x w_2$ correspond to the right one. $\lambda = \exp(\mathrm{i}kL)$ stands for the propagation constant with k being the wavenumber.

Now inserting Eq. (1) into Eqs. (3) and (4) we have

$$\begin{bmatrix} -\lambda & 0 & 1 & 0 \\ 0 & -\lambda & 0 & 1 \end{bmatrix} \begin{bmatrix} \boldsymbol{\varphi}_2^\top \\ \partial_x \boldsymbol{\varphi}_2^\top \\ \boldsymbol{\varphi}_1^\top \\ \partial_x \boldsymbol{\varphi}_1^\top \end{bmatrix} \boldsymbol{A} = \begin{bmatrix} 0 \\ 0 \end{bmatrix} =: \tilde{\boldsymbol{\Lambda}}\boldsymbol{\Phi}^\top\boldsymbol{A} = \boldsymbol{0}, \tag{5}$$

where we have identified $\tilde{\boldsymbol{\Lambda}}$ and $\boldsymbol{\Phi}$.

The key step here is to attain the nullspace basis prescribed by Eq. (5). With the indication in Ref. [29], we can get the basis $\boldsymbol{Z}$ for $\mathcal{N}\left(\tilde{\boldsymbol{\Lambda}}\boldsymbol{\Phi}^\top\right)$,

$$\boldsymbol{Z} = \left[\boldsymbol{Z}_\varphi, \mathcal{C}_{\boldsymbol{Z}}\right] = \left[\boldsymbol{Z}_\varphi, \mathcal{C}\right]\begin{bmatrix} \boldsymbol{I} & \boldsymbol{0} \\ \boldsymbol{0} & \boldsymbol{Z}_r \end{bmatrix}, \tag{6}$$

where

$$\boldsymbol{Z}_r := \begin{bmatrix} 1 & 0 \\ 0 & 1 \\ \lambda & 0 \\ 0 & \lambda \end{bmatrix} = \begin{bmatrix} \boldsymbol{I} \\ \lambda\boldsymbol{I} \end{bmatrix}, \tag{7}$$

$$\boldsymbol{Z}_{\varphi} = \mathcal{N}\left(\boldsymbol{\Phi}^{\mathsf{T}}\right), \tag{8}$$

and

$$\mathbf{C} = \left(\boldsymbol{\Phi}^{\mathsf{T}}\right)^{+}. \tag{9}$$

Assuming the final response $\boldsymbol{A}$ is linearly constituted by the basis of $\boldsymbol{Z}$, that is, $\boldsymbol{A} = \boldsymbol{Z}\boldsymbol{k}$, we arrive at

$$\left(\boldsymbol{K} - \omega^2\boldsymbol{M}\right)\boldsymbol{Z}\boldsymbol{k} = \mathbf{0}, \tag{10}$$

with $\boldsymbol{k}$ being the unknown vector to be solved. Eq. (10) becomes the equation of motion satisfying the periodic boundary conditions in Eqs. (3) and (4).

To simplify the computation, we define $\boldsymbol{Z}_c \equiv \left[\boldsymbol{Z}_{\varphi}, \mathcal{C}\right]$ and

$$\boldsymbol{Z}_l \equiv \begin{bmatrix} 1 & 0 & \lambda^{-1} & 0 \\ 0 & 1 & 0 & \lambda^{-1} \end{bmatrix} \equiv \left[\boldsymbol{I}\lambda^{-1}\ \boldsymbol{I}\right]. \tag{11}$$

Pre-multiplying Eq. (10) by $\text{diag}(\boldsymbol{I}, \boldsymbol{Z}_l)\boldsymbol{Z}_c^{\mathsf{T}}$ we get

$$\begin{bmatrix} \boldsymbol{I} & \mathbf{0} \\ \mathbf{0} & \boldsymbol{Z}_l \end{bmatrix} \left(\overline{\boldsymbol{K}} - \omega^2\overline{\boldsymbol{M}}\right) \begin{bmatrix} \boldsymbol{I} & \mathbf{0} \\ \mathbf{0} & \boldsymbol{Z}_r \end{bmatrix} \boldsymbol{k} = \mathbf{0}, \tag{12}$$

where $\overline{\boldsymbol{K}}_{n\times n}{:=}\boldsymbol{Z}_c^{\mathsf{T}}\boldsymbol{K}\boldsymbol{Z}_c$ and $\overline{\boldsymbol{M}}_{n\times n}{:=}\boldsymbol{Z}_c^{\mathsf{T}}\boldsymbol{M}\boldsymbol{Z}_c$ are square matrices.

Now, it is possible to compute the real dispersion curves once a wavenumber k is given. However, our purpose is to get the complex dispersion curves, where ω is fixed and k is to be determined. To do so, we define $\boldsymbol{D} \equiv \overline{\boldsymbol{K}} - \omega^2\overline{\boldsymbol{M}}$ and partition it as

$$\boldsymbol{D}_{n\times n} = \begin{bmatrix} \boldsymbol{D}_{11,(n-4)\times(n-4)} & \boldsymbol{D}_{12,(n-4)\times 4} \\ \boldsymbol{D}_{12,4\times(n-4)}^{\mathsf{T}} & \boldsymbol{D}_{22,4\times 4} \end{bmatrix}. \tag{13}$$

Now, Eq. (12) becomes

$$\begin{aligned} \boldsymbol{D}_{11}\boldsymbol{k}_1 + \boldsymbol{D}_{12}\boldsymbol{Z}_r\boldsymbol{k}_2 &= \mathbf{0}, \\ \boldsymbol{Z}_l\boldsymbol{D}_{12}^{\mathsf{T}}\boldsymbol{k}_1 + \boldsymbol{Z}_l\boldsymbol{D}_{22}\boldsymbol{Z}_r\boldsymbol{k}_2 &= \mathbf{0}, \end{aligned} \tag{14}$$

where $\left[\boldsymbol{k}_1^{\mathsf{T}}, \boldsymbol{k}_2^{\mathsf{T}}\right] \equiv \boldsymbol{k}$. Clearing $\boldsymbol{k}_1$ we have

$$\boldsymbol{Z}_l\left[\boldsymbol{D}_{22} - \boldsymbol{D}_{12}^{\mathsf{T}}\boldsymbol{D}_{11}^{-1}\boldsymbol{D}_{12}\right]\boldsymbol{Z}_r\boldsymbol{k}_2 = \mathbf{0}. \tag{15}$$

The Schur complement of $\boldsymbol{D}_{11}$ can be designated as $\boldsymbol{G}{:=}\boldsymbol{D}_{22} - \boldsymbol{D}_{12}^{\mathsf{T}}\boldsymbol{D}_{11}^{-1}\boldsymbol{D}_{12}$, which can be further partitioned as

$$\boldsymbol{G} = \begin{bmatrix} \boldsymbol{G}_{11} & \boldsymbol{G}_{12} \\ \boldsymbol{G}_{12}^{\mathsf{T}} & \boldsymbol{G}_{22} \end{bmatrix}. \tag{16}$$

Recalling the definition of $\boldsymbol{Z}_l$ and $\boldsymbol{Z}_r$, we can obtain the following equation,

$$\begin{bmatrix} I & \lambda^{-1} I \end{bmatrix} \begin{bmatrix} G_{11} & G_{12} \\ G_{12}^{\mathrm{T}} & G_{22} \end{bmatrix} \begin{bmatrix} I \\ \lambda I \end{bmatrix} k_2 = \left(G_{11} + G_{22} + G_{12}^{\mathrm{T}} \lambda^{-1} + G_{12} \lambda \right) k_2 = \mathbf{0}, \tag{17}$$

which can be converted into a quadratic eigenvalue problem for λ. Therefore, the wavenumber $k \in \mathbb{C}$ can be solved for a given $\omega \in \mathbb{R}$.

3. Numerical results

Once built the computational model, we start to show some numerical results. As shown in **Figure 1**, the lattice constant is selected as $L = 0.1$ m, and the beam thickness at the uniform part is $h_{uni} = 3$ mm. The ABH profile is defined as $h(x) = \varepsilon x^m + h_c$, where $m = 2$, $h_c = 0.6$ mm, $\varepsilon = (h_{uni} - h_c)/r_{abh}^m = 6\,\mathrm{m}^{-1}$, and $r_{abh} = 2$ cm. The beam is made of steel, with Young modulus being $E = 210$ GPa, density $\rho = 7800$ kg/m^3, and loss factor $\eta = 0.005$. On the other hand, the damping layer has size $r_d = 2$ cm and $h_d = 1.8$ mm, and the Young modulus $E_d = 5$ GPa, density $\rho_d = 950$ kg/m^3, and loss factor $\eta_d = 0.5$.

In the following, we will first examine the complex dispersion curves, then the transmissibilities are inspected to ensure the existence of topological low-frequency bangap.

3.1 Complex dispersion curves

To start with, we do not consider any damping (no damping layer and $\eta = 0$), and a standard periodic cell is considered. As illustrated in **Figure 2a** and **b**, where the real and imaginary parts of the complex dispersion curves have been computed. It is observed that there are two bandgaps appearing in 175–555 Hz and 1635–2330 Hz, respectively. According to the analysis in Ref. [10], those bandgaps are induced by the Bragg scattering effect. Now, we take the supercell (as depicted in **Figure 1b**), with $d = \alpha L$, where $\alpha = 1$. Please note that in this case the bandgaps should occur at the same frequency range. As demonstrated in **Figure 2c** and **d**, this is true, indicating that our WRRM is accurate to recover the complex dispersion curves. Particularly, in **Figure 2d** we found that the evanescent wave is doubled compared to that in **Figure 2b**. This is reasonable because two cells provide more wave attenuation. Please note that there are two passbands and two stopbands in the frequency range of interest (see **Figure 2a**), computed from the periodic cells. However, when using supercell structure, each passband has been divided into two ones (see $f = 50$ Hz and $f = 1040$ Hz), which gives us the possibility to open them in the following content.

Now, we shrink the distance of the ABHs in a supercell (see **Figure 1c**), and set $\alpha = 0.5$. The complex dispersion curves are illustrated in **Figure 3**. It is seen that the passbands are opened at $f = 50$ Hz and $f = 1040$ Hz, and there emerge two new bandgaps in 45–80 Hz and 895–1100 Hz. Since we are more interested in low frequencies, the new first bandgap (45–80 Hz) is much lower than the former one (175–555 Hz). Particularly, we have only changed the ABH distance and the structure has the same weight as before.

At this stage, we include the damping layers and the intrinsic loss of the beam ($\eta = 0.005$). The complex dispersion curves of a supercell with $\alpha = 1$ have been demonstrated in **Figure 4a** and **b**, which are nothing but corresponding to standard periodic lattice. In such a case we see that *bandgaps* no longer exist in the real dispersion curves but in the imaginary part (see **Figure 4b**). Also, in the passbands we also have some wave attenuation because of damping (again see **Figure 4b**).

Once we decrease the ABH distance and set $\alpha = 0.5$ (see **Figure 4c** and **d**), as expected, we can get a new bandgap at very low frequency $f = 95$ Hz. Please note that at very high frequencies the ABH effect starts to take control, so the imaginary part is very smooth and steadily becomes larger over $f = 2000$ Hz.

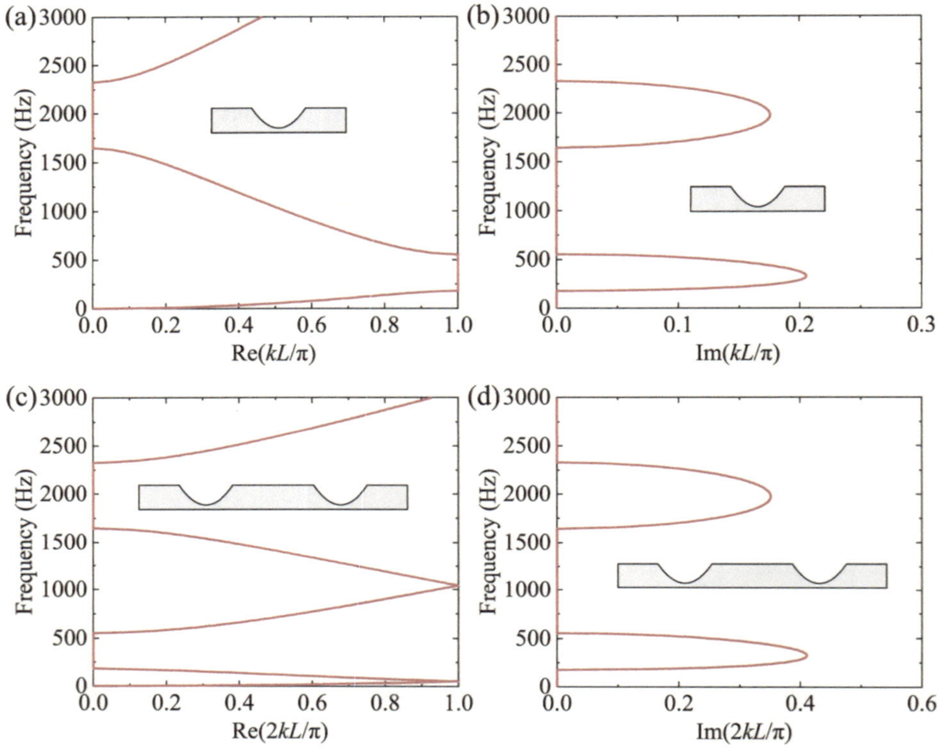

Figure 2.
(a) Real and (b) imaginary part of the complex dispersion curves of a standard periodic ABH cell. (c) Real and (d) imaginary part of the complex dispersion curves of a standard periodic supercell. Please note that in this case we have not included the damping layer and beam intrinsic loss.

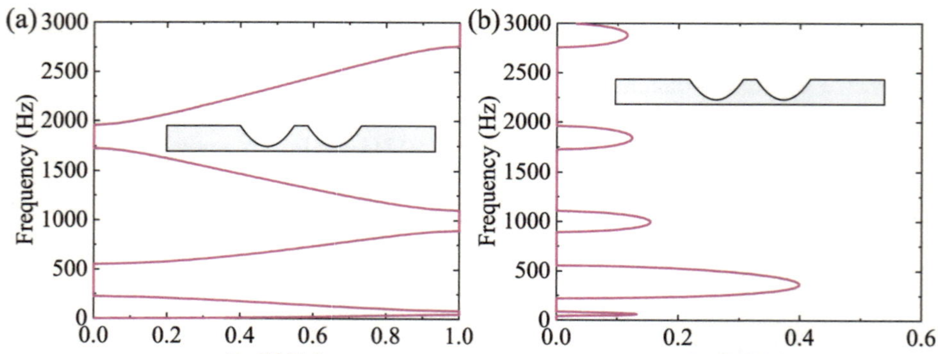

Figure 3.
(a) Real and (b) imaginary part of the complex dispersion curves of a topological supercell with $\alpha = 0.5$. Please note that in this case we have not included the damping layer and beam intrinsic loss.

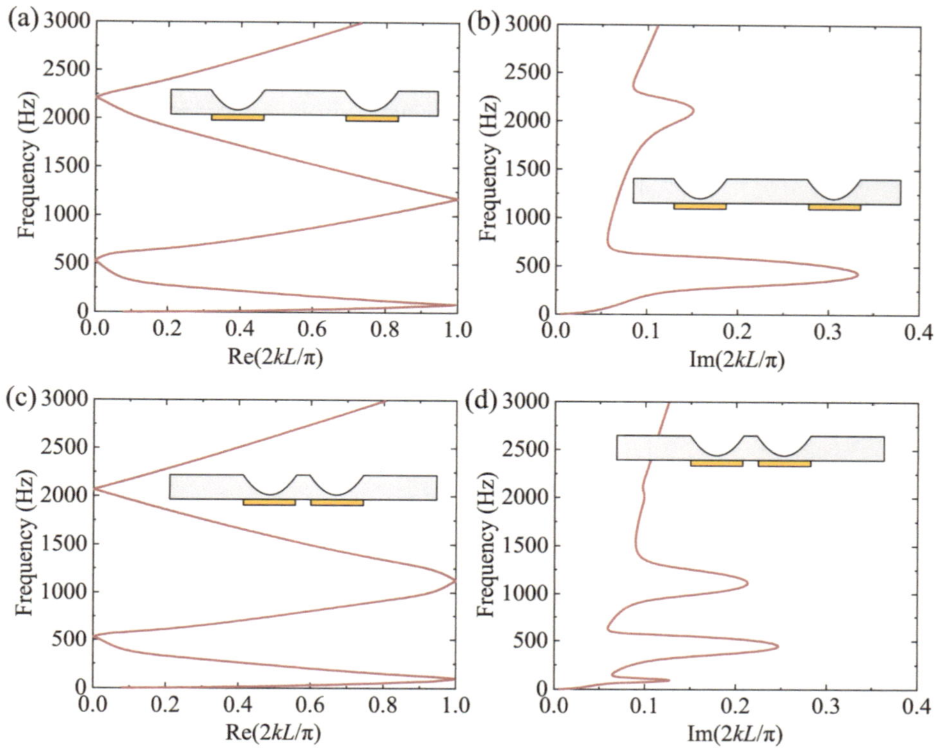

Figure 4.
(a) Real and (b) imaginary part of the complex dispersion curves of a standard periodic supercell. (c) Real and (d) imaginary part of the complex dispersion curves of a topological supercell with $\alpha = 0.5$. Please note that in this case we have included the damping layer and beam intrinsic loss.

3.2 Transmissibilities

We next shed light on the wave transmission in finite beams. As usual, we first exclude any damping. In total, 12 cells have been taken into account; thus, the beam

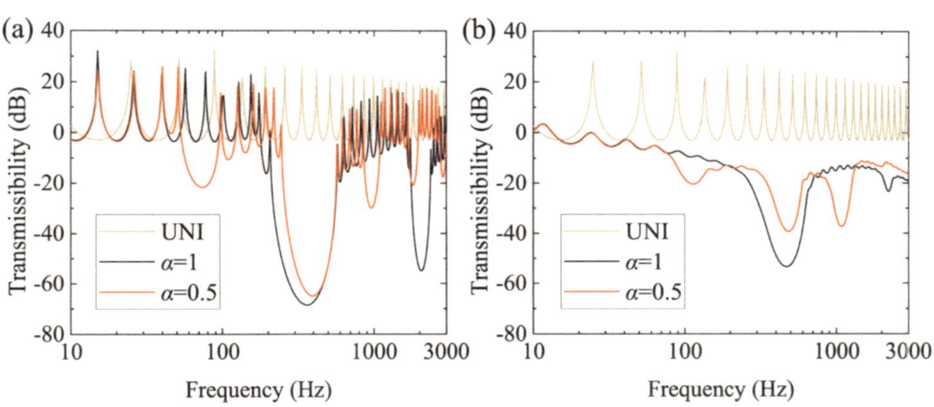

Figure 5.
Transmissibilities of finite beams with 12 cells. (a) Without damping layer nor the beam intrinsic loss. (b) With damping layer and the beam intrinsic loss.

length equals to 1.2 m. For reference, we have computed the transmissibillity, $T = 10\lg[w^2(1.2)/w^2(0)]$, of a uniform beam (UNI) where a point force is exerted at $x = 0$, see the brown line in **Figure 5a**. For the better inspection of low frequencies, we have chosen logarithmic scale for the frequency axis. Not only that, the T of standard periodic ABH beam has also been computed, see the black line in **Figure 5a**. We can find that two bandgaps take places at the frequency ranges predicted by the dispersion curves in **Figure 2**. Once we have set $\alpha = 0.5$ (see the red curve in **Figure 5a**), we can get a new bandgap centred at 70 Hz.

Once the damping layer and intrinsic loss $\eta = 0.005$ are taken into account, it is seen that vibration peaks are substantially smoothed, especially at high frequencies, where the ABH effect starts working. However, the most intriguing phenomenon is the presence of the low-frequency topological bandgap at $f = 100$ Hz, confirming that the proposed supercell ABH with shrinking distance is very promising for low-frequency vibration reduction.

4. Conclusions

In this chapter, we have proposed a topological ABH metamaterial to generalize a low-frequency bandgap for vibration reduction. Based on a standard periodic ABH beam, we have taken a supercell and reduced the ABH distance, opening new bandgaps, especially at low frequencies.

To characterize such a feature, we have adopted the wave and Rayleigh-Ritz method (WRRM) to compute the complex dispersion curves, which relies on solving the quadratic eigenvalue problem of $k \in \mathbb{C}$ for a fixed $\omega \in \mathbb{R}$. After that, we have analyzed impact of changing the ABH distance on the bandgaps. Before including the damping, the bandgaps can be both appreciated in the real and imaginary parts of the complex dispersion curves. However, after considering the damping, bandgaps can only be inspected in the imaginary part.

Both the analyses on the dispersion curves and tranmissibilities indicate that the topological ABHs can indeed produce a low-frequency bandgap, which constitutes a promising low-frequency wave manipulation technique.

Acknowledgements

J. Deng acknowledges the support received by the National Natural Science Foundation of China (52301386) and the support of the Beatriu de Pins postdoctoral program of the Department of Research and Universities of the Generalitat of Catalonia (2022 BP 00027).

Conflict of interest

The author declares no conflict of interest.

Author details

Jie Deng[1,2]

1 Chongqing Industry Polytechnic College, Chongqing, China

2 Human-Environment Research (HER) Group, Department of Engineering, La Salle, Universitat Ramon Llull C/Quatre Camins, Barcelona, Catalonia, Spain

*Address all correspondence to: dengjie_cn@outlook.com

References

[1] Huang W, Tao C, Ji H, Qiu J. Enhancement of wave energy dissipation in two-dimensional acoustic black hole by simultaneous optimization of profile and damping layer. Journal of Sound and Vibration. 2021;**491**:115764

[2] Chen X, Jing Y, Zhao J, Deng J, Cao X, Pu H, et al. Tunable shunting periodic acoustic black holes for low-frequency and broadband vibration suppression. Journal of Sound and Vibration. 2024; **580**:118384

[3] Bu Y, Tang Y, Ding Q. Novel vibration self-suppression of periodic pipes conveying fluid based on acoustic black hole effect. Journal of Sound and Vibration. 2023;**567**:118077

[4] Sheng H, He MX, Zhao J, Kam CT, Ding Q, Lee HP. The ABH-based lattice structure for load bearing and vibration suppression. International Journal of Mechanical Sciences. 2023;**252**:108378

[5] Gao W, Qin Z, Chu F. Broadband vibration suppression of rainbow metamaterials with acoustic black hole. International Journal of Mechanical Sciences. 2022;**228**:107485

[6] Zhu H, Semperlotti F. Phononic thin plates with embedded acoustic black holes. Physical Review B. 2015;**91**(10): 104304

[7] Zhu H, Semperlotti F. Anomalous refraction of acoustic guided waves in solids with geometrically tapered metasurfaces. Physical Review Letters. 2016;**117**(3):034302

[8] Zhu H, Semperlotti F. Two-dimensional structure-embedded acoustic lenses based on periodic acoustic black holes. Journal of Applied Physics. 2017;**122**(6):065104

[9] Tang L, Cheng L. Broadband locally resonant band gaps in periodic beam structures with embedded acoustic black holes. Journal of Applied Physics. 2017; **121**(19):194901. DOI: 10.1063/1.4983459

[10] Deng J, Guasch O. On the bandgap mechanism of periodic acoustic black holes. Journal of Sound and Vibration. 2024;**579**:118379

[11] Tang L, Cheng L. Ultrawide band gaps in beams with double-leaf acoustic black hole indentations. The Journal of the Acoustical Society of America. 2017; **142**(5):2802-2807

[12] Zhang Y, Chen K, Zhou S, Wei Z. An ultralight phononic beam with a broad low-frequency band gap using the complex lattice of acoustic black holes. Applied Physics Express. 2019;**12**(7): 077002

[13] Gao N, Guo X, Deng J, Cheng B, Hou H. Elastic wave modulation of double-leaf ABH beam embedded mass oscillator. Applied Acoustics. 2021;**173**: 107694

[14] Park S, Jeon W. Ultra-wide low-frequency band gap in a tapered phononic beam. Journal of Sound and Vibration. 2021;**499**:115977

[15] Tang L, Cheng L, Chen K. Complete sub-wavelength flexural wave band gaps in plates with periodic acoustic black holes. Journal of Sound and Vibration. 2021;**502**:116102

[16] Deng J, Zheng L, Gao N. Broad band gaps for flexural wave manipulation in plates with embedded periodic strip acoustic black holes. International Journal of Solids and Structures. 2021; **224**:111043. DOI: 10.1016/j.ijsolstr.2021.111043

[17] Deng J, Guasch O, Maxit L, Gao N. A metamaterial consisting of an acoustic black hole plate with local resonators for broadband vibration reduction. Journal of Sound and Vibration. 2022;**526**: 116803

[18] Deng J, Guasch O, Maxit L, Gao N. Sound radiation and non-negative intensity of a metaplate consisting of an acoustic black hole plus local resonators. Composite Structures. 2023;**304**:116423

[19] Li H, O'donoughue P, Masson F, Pelat A, Gautier F, Touzé C. Broadband shock vibration absorber based on vibro-impacts and acoustic black hole effect. International Journal of Non-Linear Mechanics. 2024;**159**:104620

[20] Deng J, Chen X, Yang Y, Qin Z, Guo W. Periodic additive acoustic black holes to absorb vibrations from plates. International Journal of Mechanical Sciences. 2024;**267**:108990

[21] Deng J, Ma J, Chen X, Yang Y, Gao N, Liu J. Vibration damping by periodic additive acoustic black holes. Journal of Sound and Vibration. 2024; **574**:118235

[22] Deng J, Gao N, Chen X. Ultrawide attenuation bands in gradient metabeams with acoustic black hole pillars. Thin-Walled Structures. 2023; **184**:110459

[23] Wang T, Tang Y, Yang T, Ma ZS, Ding Q. Bistable enhanced passive absorber based on integration of nonlinear energy sink with acoustic black hole beam. Journal of Sound and Vibration. 2023;**544**:117409

[24] Ganti SS, Liu TW, Semperlotti F. Topological edge states in phononic plates with embedded acoustic black holes. Journal of Sound and Vibration. 2020;**466**:115060

[25] Lyu X, Li H, Ma Z, Ding Q, Yang T, Chen L, et al. Numerical and experimental evidence of topological interface state in a periodic acoustic black hole. Journal of Sound and Vibration. 2021;**514**:116432

[26] Deng J, Xu Y, Guasch O, Gao N, Tang L, Chen X. A two-dimensional wave and Rayleigh–Ritz method for complex dispersion in periodic arrays of circular damped acoustic black holes. Mechanical Systems and Signal Processing. 2023;**200**:110507

[27] Deng J, Zheng L, Zeng P, Zuo Y, Guasch O. Passive constrained viscoelastic layers to improve the efficiency of truncated acoustic black holes in beams. Mechanical Systems and Signal Processing. 2019;**118**:461-476. DOI: 10.1016/j.ymssp.2018.08.053

[28] Deng J, Zheng L, Guasch O, Wu H, Zeng P, Zuo Y. Gaussian expansion for the vibration analysis of plates with multiple acoustic black holes indentations. Mechanical Systems and Signal Processing. 2019;**131**:317-334. DOI: 10.1016/j.ymssp.2019.05.024

[29] Deng J, Xu Y, Guasch O, Gao N, Tang L, Guo W. A wave and Rayleigh–Ritz method to compute complex dispersion curves in periodic lossy acoustic black holes. Journal of Sound and Vibration. 2023;**546**:117449

Chapter 5

Nonlinear Phononics and Coupled Mode Theory

Ali Rostami and Hodjat Ahmadi

Abstract

Understanding phononic wave propagation in nonlinear and coupled media is inevitable in designing devices to harness phonons. In photonics, the nonlinear Schrodinger equation (NSE) and coupled mode theory (CMT) are extensively used to manipulate the propagation of electromagnetic (EM) waves. Since phonons as the quasi-particles related to mechanical vibrations are similar to their photonic counterparts (photons), inspired by EM, the NSE and CMT are developed for elastic or phononic waves. In the first section, from the novel point of view, the nonlinearity is dealt with and consequently applied to the equation of motion to derive the NSE. In the following, a hard limiter is introduced as an application of nonlinear wave propagation. The coupled-mode equations are obtained for the elastic slab waveguides in the second section. These equations are applied to explain the wave propagation in two parallel and nonparallel waveguides.

Keywords: nonlinear Schrodinger equation, hard limiter, coupled mode theory, slab waveguide, shear waves, Rayleigh-Lame waves

1. Introduction

Phonons, the quantas of mechanical vibration, play a significant role not only in phononic systems like phononic crystals but also in optoelectronic and optomechanical devices where electron-phonon and photon-phonon interactions have an undeniable impact on the operation of devices. Furthermore, they can influence the operation of photonic devices and nano and micro-electromechanical systems (N/MEMS). Raman and Brillouin scattering are the most famous phonon-involved phenomena. Cavity optomechanics, superconducting circuits, cell manipulation, quantum acoustics, and communication devices are some hot topic issues utilizing phononic waves [1]. The electron-phonon interaction affects the mobility of carriers in semiconductors, the electrical resistance of metals, carrier density in graphene, and optical absorption of the indirect semiconductors [2, 3]. Moreover, the phonon-photon interaction can be manipulated to change the group velocity used in the integrated quantum optomechanical memories and signal processing [4, 5]. Due to these numerous applications and effects, harnessing phonons is of interest to engineers and physicists. The nonlinear media and coupled mode theory are significant in controlling wave propagation regardless of its type, which can be electronic, photonic, or phononic. This book chapter is generally classified into nonlinear phononics and coupled mode theory.

2. Nonlinear phononics

Nonlinearity has broad applications in electronics and photonics. Designing Schmitt triggers in electronics, full-optical analog to digital converters, and optical microscopy are some examples of the use of nonlinearity [6–10]. Generally, a nonlinear medium is referred to as a medium in which the basic parameters depend on the amplitude of the transmitted signal. The nonlinear media provide the possibility of controlling wave propagation via other waves. In electromagnetics and photonics, the nonlinear Schrödinger equation (NSE) was introduced and has been extensively utilized to investigate nonlinear phenomena such as second harmonic generation, four-wave mixing, and modulation instability [11]. In this section, The NSE is introduced for phononic or elastic waves in an elastic medium. Furthermore, this equation is applied to describe the operation of a hard limiter in which the amplitude of the output wave is confined to the desired value.

2.1 Elastic waves

As the electromagnetic wave equation describes the optical wave propagation, the phononic waves in solids are governed by the equation of motion given by Eq. (1) in which ρ, u_i and σ_{ij} are mass density, displacement, and stress tensor [12].

$$\rho \ddot{u}_i = \frac{\partial \sigma_{ij}}{\partial x_j} \tag{1}$$

The stress tensor is related to the strain tensor via Eq. (2), where c_{ijkl} is the stiffness tensor and e_{kl} is the strain tensor defined by Eq. (3) [12].

$$\sigma_{ij} = c_{ijkl} e_{kl} \tag{2}$$

$$e_{ij} = \frac{1}{2}\left(\frac{\partial u_i}{\partial x_j} + \frac{\partial u_j}{\partial x_i}\right) \tag{3}$$

Using Voigt notation, the stiffness tensor for cubic elasticity is expressed by a 6×6 matrix in which λ and μ are Lamb parameters [12].

$$c_{ij} = \begin{pmatrix} \lambda + 2\mu & \lambda & 0 & 0 & 0 & 0 \\ \lambda & \lambda + 2\mu & \lambda & 0 & 0 & 0 \\ \lambda & \lambda + 2\mu & \lambda & 0 & 0 & 0 \\ 0 & 0 & 0 & \lambda & 0 & 0 \\ 0 & 0 & 0 & 0 & \lambda & 0 \\ 0 & 0 & 0 & 0 & 0 & \lambda \end{pmatrix} \tag{4}$$

The wave equation for the displacement is obtained using Eqs. (1) to (4) and given by [13]:

$$\mu \nabla^2 \vec{u} + (\lambda + \mu) \nabla \left(\nabla . \vec{u} \right) = \rho \frac{\partial^2 \vec{u}}{\partial t^2} \tag{5}$$

2.2 Mass density as the fundamental parameter of nonlinearity

Lennard-Jones potential, shown in **Figure 1**, is a simple mathematical model for the interaction between neutral atoms and molecules. According to the mass-spring model, an atom located in the equilibrium point with the minimum potential energy is considered a particle with a certain mass connected to the fixed particle at the origin through a spring. Around this equilibrium point, the potential can be approximated quadratically; hence, the force, which is the gradient of potential energy, named by Hooke's law, is linear. Increasing the force on the particle causes increasing displacement; hence, more force changes the particle's position more than before. Therefore, the quadratic approximation will no longer be valid. It leads to the nonlinear relationship between displacement and force. The greater the displacement from the equilibrium distance, the more the compression or expansion of the spring. Consequently, the total mass in unit length is increased or decreased due to the flux of the displacement field [14].

2.2.1 Mathematical modeling of nonlinearity

The deformation is defined as Eq. (6), where ΔV, V_0, and u are volume change, initial volume, and displacement, respectively [14].

$$\frac{\Delta V}{V_0} = \nabla . \vec{u} \tag{6}$$

Using the conservation of mass and Eq. (6), it can be obtained:

$$\rho_0 V_0 = \rho V = \rho V_0 \left(1 + \nabla . \vec{u} \right) \tag{7}$$

where $V = V_0 + \Delta V$ and ρ is the perturbed mass density. Therefore ρ is expressed as:

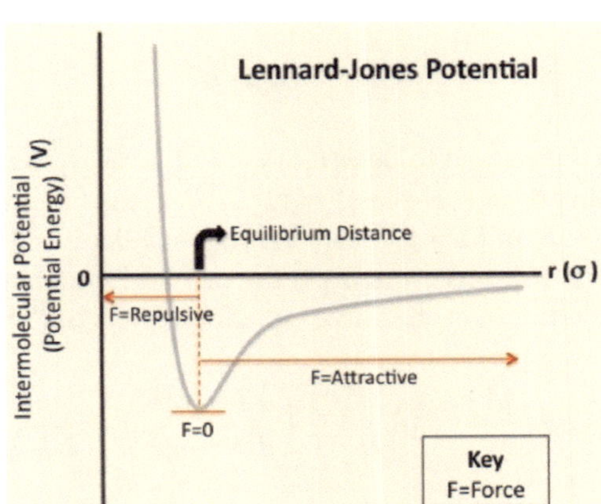

Figure 1.
Lennard-Jones potential energy.

$$\rho = \frac{\rho_0}{1 + \nabla.\vec{u}} \tag{8}$$

Since the deformation is small compared to the initial volume, the perturbed mass density can be described as:

$$\rho = \rho_0 \left(1 - \nabla.\vec{u} + \left(\nabla.\vec{u}\right)^2 + \cdots\right) \tag{9}$$

The displacement can be represented by scalar (φ) and vector (ψ) potential as the following equation [13, 15, 16].

$$u = \nabla\varphi + \nabla \times \vec{\psi} \tag{10}$$

Applying Eq. (10) to Eq. (5) leads to the wave equations for scalar and vector potential [13, 16].

$$\nabla^2\varphi = \left(\frac{\rho_0}{\lambda + 2\mu}\right)\frac{\partial^2\varphi}{\partial t^2} \tag{11}$$

$$\nabla^2\vec{\psi} = \left(\frac{\rho_0}{\mu}\right)\frac{\partial^2\vec{\psi}}{\partial t^2} \tag{12}$$

Eq. (10) indicates that $\nabla.\vec{u} = \nabla^2\varphi$ and hence, Eq. (9) can be expressed by scalar potential as:

$$\rho = \rho_0\left(1 - \nabla^2\varphi + \left(\nabla^2\varphi\right)^2 + \cdots\right) \tag{13}$$

2.3 Nonlinear schrödinger equation (NSE)

The wave equation for scalar potential is expressed as Eq. (13), in which $n_0 = \sqrt{\frac{\rho_0}{\lambda+2\mu}}$ is considered the refractive index similar to EM waves [14].

$$\nabla^2\varphi = -n_0^2\omega^2\varphi \tag{14}$$

The scalar potential is assumed to be $\varphi(r,t) = (1/2)\{F(x,y)A(z,t)\exp[i(\beta_0 z - \omega t)] + c.c\}$ in the time domain and $\varphi(r, \omega - \omega_0) = (1/2)\{F(x,y)A(z, \omega - \omega_0)\exp i(\beta_0 z) + c.c\}$ in the frequency domain. A and F are slowly varying envelope and mode shapes, respectively. β_0 is the wave number and ω_0 is the central frequency of the input pulse. Inserting this assumed response in the wave equation results in [14]:

$$\frac{1}{A}\left(\frac{\partial^2 A}{\partial z^2} + 2j\beta_0\frac{\partial A}{\partial z} - \beta_0^2 A\right) + \frac{1}{F}\left(\frac{\partial^2 F}{\partial x^2} + \frac{\partial^2 F}{\partial y^2}\right) + \frac{\rho_0}{\lambda + 2\mu}\omega^2 = 0 \tag{15}$$

Two eigenvalue equations for A and F (Eqs. 16 and 17) are obtained using the separation of variable method. In Eq. (15), the second derivation of amplitude is neglected because of its slow variation.

$$\frac{\partial^2 F}{\partial x^2} + \frac{\partial^2 F}{\partial y^2} + \left(\frac{\rho_0}{\lambda + 2\mu}\omega^2 - \tilde{\beta}^2\right)F = 0 \tag{16}$$

$$2j\beta_0 \frac{\partial A}{\partial z} + \left(\tilde{\beta}^2 - \beta_0^2\right)A = 0 \tag{17}$$

The perturbed mass density (ρ) equals $\rho_0\left(1 - \nabla^2\varphi\right)$ in the first order perturbation. This mass density causes the refractive index to be perturbed:

$$n = n_0 + \Delta n \triangleq \sqrt{\frac{\rho_0 + \Delta\rho}{\lambda + 2\mu}} \tag{18}$$

As $\Delta\rho \ll \rho_0$ the perturbation in the refractive index is obtained by Eq. (19), which changes the eigenvalue ($\tilde{\beta}$) (Eq. 20). The value of this alteration is calculated by Eq. (21) [11].

$$\Delta n = -\frac{1}{2}\left(\frac{\rho_0}{\lambda + 2\mu}\right)^{\frac{3}{2}} \omega_0^2 |\varphi| \tag{19}$$

$$\tilde{\beta}(\omega) = \beta(\omega) + \Delta\beta \tag{20}$$

$$\Delta\beta = -\frac{1}{2}\left(\frac{\rho_0}{\lambda + 2\mu}\right)^{\frac{3}{2}} \omega_0^3 \frac{1}{A_{eff}} |A| \tag{21}$$

A_{eff} is defined as Eq. (22). It is the effective area where most wave energy is confined.

$$A_{eff} = \frac{\left(\int\limits_{-\infty}^{+\infty}\int\limits_{-\infty}^{+\infty} |F|^2 dxdy\right)^2}{\int\limits_{-\infty}^{+\infty}\int\limits_{-\infty}^{+\infty} |F|^3 dxdy} \tag{22}$$

The eigenvalue equation for the amplitude is written as Eq. (23) using Eq. (21).

$$\frac{\partial A}{\partial z} = j(\beta(\omega) + \Delta\beta - \beta_0)A \tag{23}$$

According to the dependence of the wave number on the angular frequency, it can be expanded around the central frequency of the input pulse in the form of Eq. (24), where $1/\beta_1$ is the group velocity.

$$\beta(\omega) = \beta_0 + (\omega - \omega_0)\beta_1 + \frac{1}{2}(\omega - \omega_0)^2\beta_2 + \frac{1}{6}(\omega - \omega_0)^3\beta_3 + \ldots \tag{24}$$

Inserting the above equation in Eq. (23) and applying the inverse Fourier transform results in the nonlinear Schrödinger equation (NSE) [14]:

$$\frac{\partial A}{\partial z} = -\left(\beta_1 \frac{\partial A}{\partial t} + i\beta_2 \frac{\partial^2 A}{\partial t^2} + \ldots\right) - i\omega_0{}^3\left(\left(\frac{\rho_0}{\lambda + 2\mu}\right)^{\frac{3}{2}}\left(\frac{1}{A_{eff}}\right)\right)|A|A + \ldots \tag{25}$$

The NSE includes both the dispersion and nonlinear effects. β_1, β_2, and higher orders β indicate the first, second, and higher order dispersions. The term containing the effective area represents nonlinearity. The NSE clearly expresses how dispersion and nonlinearity affect wave amplitude in a specific position during wave propagation.

2.4 Hard limiter

NSE can be utilized to describe nonlinear phenomena such as wave mixing and phase modulation. Wave mixing is used to design phononic limiters. A limiter confines the output to a particular value when the input exceeds a threshold value. As the mass density is dynamically changed due to the wave propagation, the wave number is considered $\beta(\omega) = \omega\hat{\rho}(\omega)$ with $\hat{\rho}(\omega) = \sqrt{\rho(\omega)/(\lambda + 2\mu)}$. $\hat{\rho}(\omega)$ is introduced as the elastic wave refractive index using duality between elastic and electromagnetic waves [17].

In the 1-dimensional structure depicted in **Figure 2**, Eq. (5) is expressed as:

$$\frac{\partial^2 u(z)}{\partial z^2} = -[\omega\hat{\rho}(\omega)]^2 u(z) \tag{26}$$

The nonlinear mass density can be written as:

$$\rho = \rho_0\left(1 - \frac{u}{c_0} + \left(\frac{u}{c_0}\right)^2 + \dots\right) \tag{27}$$

where c_0 is the lattice constant in the direction of displacement. The refractive index in Eq. (26) for the structure of **Figure 2** can be expressed as [18]:

$$\begin{aligned}[\hat{\rho}(z)]^2 = {} & \hat{\rho}_{av}^2 + 2\hat{\rho}_{av}(\hat{\rho}_1 - \hat{\rho}_2)f(z) + i2\hat{\rho}_{av}[\kappa_0 + (\kappa_1 - \kappa_2)f(z)] \\ & + 2\hat{\rho}_{av}[\hat{\rho}_{NL} + (\hat{\rho}_{1NL} - \hat{\rho}_{2NL})f(z)]|u|\end{aligned} \tag{28}$$

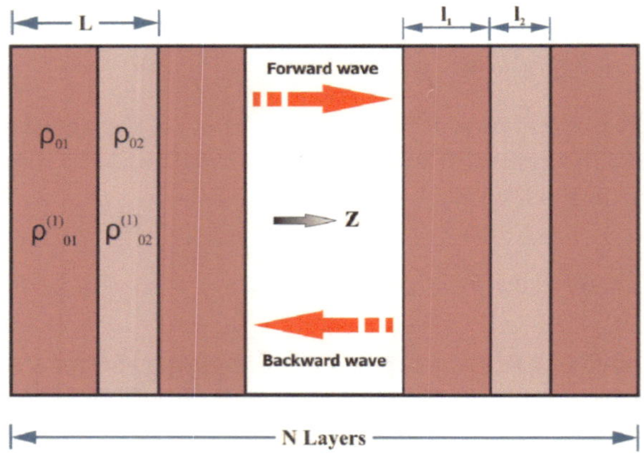

Figure 2.
The 1D periodic structure includes two types of layers with lengths of l_1 and l_2. ρ_o and $\rho_o^{(1)}$ are the linear and first nonlinear parts of mass density, respectively [14].

where $\hat{\rho}_1 = \sqrt{\rho_{01}/(\lambda_1 + 2\mu_1)}$ and $\hat{\rho}_2 = \sqrt{\rho_{02}/(\lambda_2 + 2\mu_2)}$ are linear refractive indices and $\hat{\rho}_{av}$ is the average refractive index given by:

$$\hat{\rho}_{av} = \left(1 - \frac{l_2}{L}\right)\hat{\rho}_1 + \left(\frac{l_2}{L}\right)\hat{\rho}_2 \tag{29}$$

$f(z)$ is the periodic function defined by Eq. (30) in one period.

$$f(z) = \begin{cases} \frac{l_2}{L}, 0 < z < l_1 \\ \frac{l_2}{L} - 1, l_1 < z < L \end{cases} \tag{30}$$

κ_1 and κ_2 are the imaginary parts of refractive indices, and κ_0 is their average. Similar to the linear parts, Eq. (31) defines the average nonlinear refractive index $\hat{\rho}_{NL}$. $\hat{\rho}_{1NL}$ and $\hat{\rho}_{2NL}$ are the nonlinear indices obtained through the Taylor expansion of the refractive index. These indices are expressed via Eq. (33) considering the nonlinear relation for mass density given by Eq. (32), [14].

$$\hat{\rho}_{Nl} = \left(1 - \frac{l_2}{L}\right)\hat{\rho}_{1NL} + \left(\frac{l_2}{L}\right)\hat{\rho}_{2NL} \tag{31}$$

$$\rho = \rho_0 + \rho_0^{(1)} u \tag{32}$$

$$\hat{\rho}_{1NL} = \frac{\rho_{01}^{(1)}}{2\sqrt{\rho_{01}(\lambda_1 + 2\mu_1)}} \tag{33}$$

$$\hat{\rho}_{2NL} = \frac{\rho_{02}^{(1)}}{2\sqrt{\rho_{02}(\lambda_2 + 2\mu_2)}} \tag{34}$$

The displacement field is the sum of forward and backward waves given by: $u(z) = A(z)e^{ikz} + B(z)e^{-ikz}$, in which $k = \omega\hat{\rho}_{av}$. Inserting this equation and Eq. (28) in Eq. (26) and using Eq. (35), which is the Fourier transform of f(z), results in Eq. (36) in which the slowly varying approximation is applied [18].

$$f(z) = -\sum_{m\neq 0} e^{im\pi l_2/L} \frac{\sin(m\pi l_2/L)}{m\pi} e^{i(2m\pi/L)z} \tag{35}$$

$$\begin{aligned} -i\left(\frac{dA(z)}{dz}e^{ikz} - \frac{dB(z)}{dz}e^{-ikz}\right) = & -\omega(\hat{\rho}_1 - \hat{\rho}_2)A(z)\sum_{m\neq 0} e^{im\pi l_2/L} \frac{\sin(m\pi l_2/L)}{m\pi} e^{i(2m\pi/L + k)z} \\ & -i\omega(\kappa_1 - \kappa_2)A(z)\sum_{m\neq 0} e^{im\pi l_2/L} \frac{\sin(m\pi l_2/L)}{m\pi} e^{i(2m\pi/L + k)z} \\ & -\omega(\hat{\rho}_{1NL} - \hat{\rho}_{2NL})|u(z)|A(z)\sum_{m\neq 0} e^{im\pi l_2/L} \frac{\sin(m\pi l_2/L)}{m\pi} e^{i(2m\pi/L + k)z} \\ & -\omega(\hat{\rho}_1 - \hat{\rho}_2)B(z)\sum_{m\neq 0} e^{im\pi l_2/L} \frac{\sin(m\pi l_2/L)}{m\pi} e^{i(2m\pi/L - k)z} \\ & -i\omega(\kappa_1 - \kappa_2)B(z) e^{im\pi l_2/L} \frac{\sin(m\pi l_2/L)}{m\pi} e^{i(2m\pi/L - k)z} \\ & -\omega(\hat{\rho}_{1NL} - \hat{\rho}_{2NL})|u(z)|B(z)\sum_{m\neq 0} e^{im\pi l_2/L} \frac{\sin(m\pi l_2/L)}{m\pi} e^{i(2m\pi/L - k)z} \\ & +\omega(i\kappa_0 + \hat{\rho}_{NL}|u|)\left(A(z)e^{ikz} + B(z)e^{-ikz}\right) \end{aligned} \tag{36}$$

Assuming the wavelength around the first stop band ($k \approx \pi/_L$), the phase-matching condition leads to the coupled equations for the amplitudes of forward and backward waves given by Eq. (36), [8, 18].

$$i\frac{dA(z)}{dz} = \omega[(\hat{\rho}_1 - \hat{\rho}_2) + (\hat{\rho}_{1NL} - \hat{\rho}_{2NL})|u(z)|]e^{\frac{i\pi l_2}{L}}\frac{\sin\left(\frac{\pi l_2}{L}\right)}{\pi}e^{i\left(\frac{2\pi}{L}-2k\right)z}B(z) - \hat{\rho}_{NL}|u(z)|A(z) \tag{37}$$

$$i\frac{dB(z)}{dz} = \omega[(\hat{\rho}_2 - \hat{\rho}_1) + (\hat{\rho}_{2NL} - \hat{\rho}_{1NL})|u(z)|]e^{\frac{-i\pi l_2}{L}}\frac{\sin\left(\frac{\pi l_2}{L}\right)}{\pi}e^{-i\left(\frac{2\pi}{L}-2k\right)z}A(z) + \hat{\rho}_{NL}|u(z)|B(z) \tag{38}$$

These equations will be solved for the lengths given by Eqs. (39) and (40).

$$l_1 = \frac{\lambda}{2\left(\hat{\rho}_1 - \hat{\rho}_2\frac{\hat{\rho}_{1NL}}{\hat{\rho}_{2NL}}\right)} \tag{39}$$

$$l_2 = \frac{\lambda}{2\left(\hat{\rho}_2 - \hat{\rho}_1\frac{\hat{\rho}_{2NL}}{\hat{\rho}_{1NL}}\right)} \tag{40}$$

The similarity of the linear characteristic of two layers in the structure of **Figure 2** means that the propagating medium is homogeneous in the absence of nonlinearity; hence, there is no reflected wave. However, the different nonlinear features of the two layers cause the medium to be inhomogeneous during wave propagation. The intensity of the reflected wave corresponds to the value of the nonlinear part of the refractive index.

The linear and nonlinear parts of the refractive indices are assumed to be $\hat{\rho}_1 = \hat{\rho}_2 = 1.5 \times 10^{-4}$, $\hat{\rho}_{1NL} = 10^5$, and $\hat{\rho}_{2NL} = -10^5$ to investigate the operation of the limiter. The transmission for different number of layers at the input intensity of 1.44 and different intensities at N = 50 are depicted in **Figure 3**. The thickness of layers is

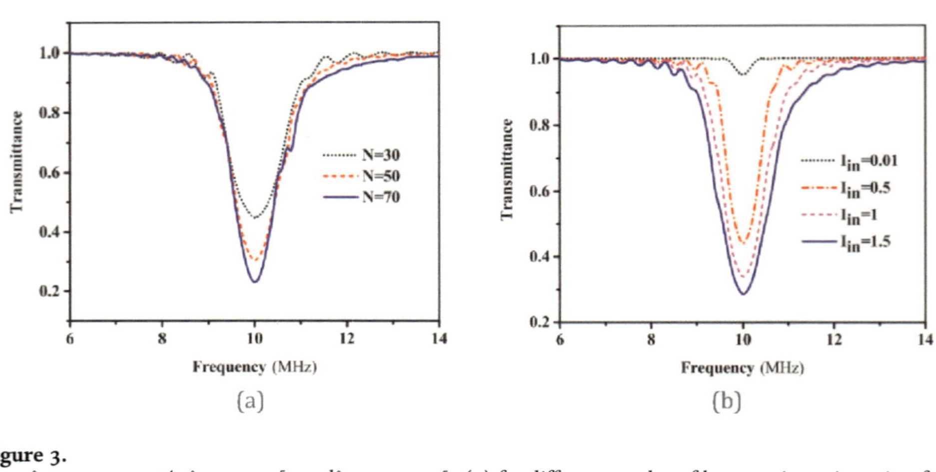

Figure 3.
$\hat{\rho}_1 = \hat{\rho}_2 = 1.5 \times 10^{-4}$, $\hat{\rho}_{1NL} = 10^5$, and $\hat{\rho}_{2NL} = -10^5$. (a) for different number of layers at input intensity of 1.44. (b) for different intensities at N = 50 [14].

set to hinder the transmission of 10 MHz. The transmission is decreased due to the increasing number of layers and intensity because of boosting the nonlinear effect [14].

Characteristic curves of the limiter are shown in **Figure 4**. **Figure 4a** illustrates that more layers result in a better limiting process. For a nonlinear index difference (NID) of 2×10^5, 100 layers are required to confine the output to 1. Furthermore, as seen in **Figure 4b**, the more the NID, the better the limiting. This figure indicates that NID = 3×10^5 is the best for 100 layers to limit the output to 1. Moreover, **Figure 4c** represents that NID can be adjusted so that the output is confined to the desirable value. In this figure, NIDs are selected so that the output is limited to 0.64, 0.81, and 1.

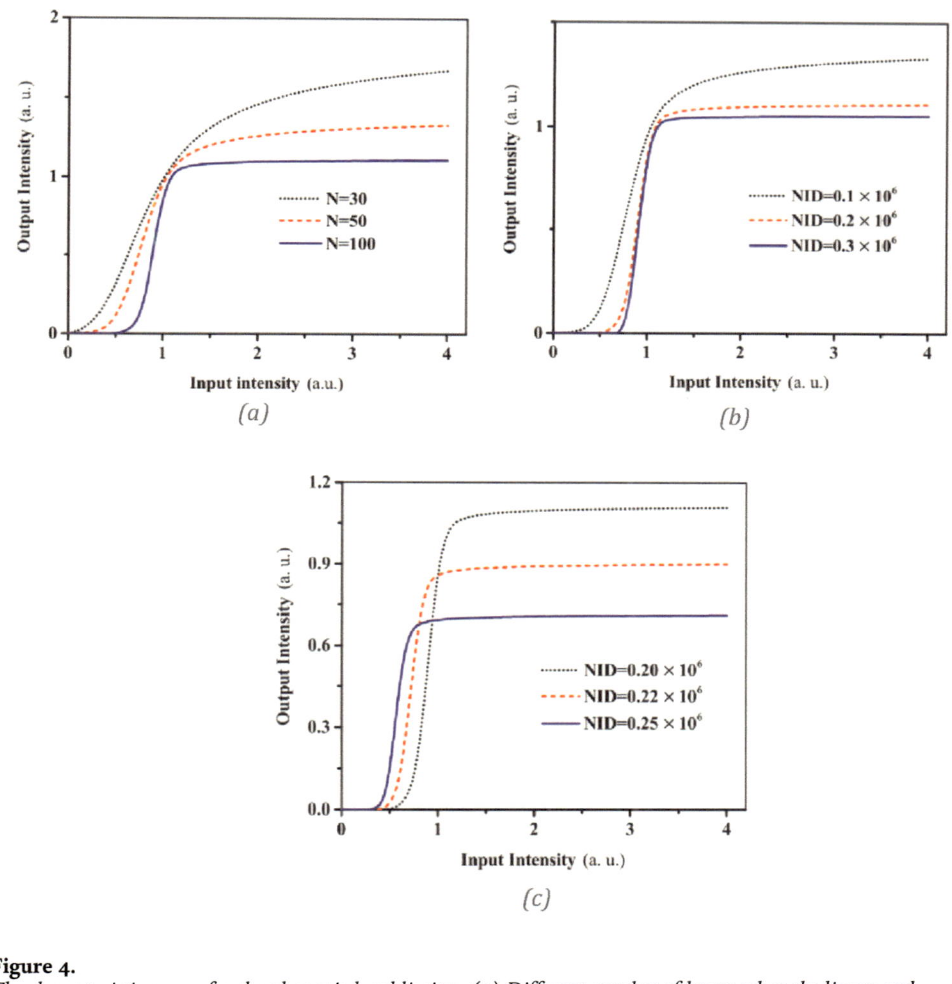

Figure 4.
The characteristic curves for the phononic hard limiter. (a) Different number of layers when the linear and nonlinear parts of the refractive index are $\hat{\rho}_1 = 1.5 \times 10^{-4}$, $\hat{\rho}_2 = 1.7 \times 10^{-4}$, $\hat{\rho}_{1NL} = 1 \times 10^5$, *and* $\hat{\rho}_{2NL} = -1 \times 10^5$. *(b) N = 100. Dotted line:* $\hat{\rho}_1 = 1.5 \times 10^{-4}$, $\hat{\rho}_2 = 1.6 \times 10^{-4}$, $\hat{\rho}_{1NL} = 0.5 \times 10^5$, *and* $\hat{\rho}_{2NL} = -0.5 \times 10^5$. *Dashed line:* $\hat{\rho}_1 = 1.5 \times 10^{-4}$, $\hat{\rho}_2 = 1.7 \times 10^{-4}$, $\hat{\rho}_{1NL} = 1 \times 10^5$, *and* $\hat{\rho}_{2NL} = -1 \times 10^5$. *Solid line:* $\hat{\rho}_1 = 1.5 \times 10^{-4}$, $\hat{\rho}_2 = 1.8 \times 10^{-4}$, $\hat{\rho}_{1NL} = 1.5 \times 10^5$, *and* $\hat{\rho}_{2NL} = -1.5 \times 10^5$. *(c) N = 100,* $\hat{\rho}_1 = 1.4 \times 10^{-4}$, *and* $\hat{\rho}_2 = 1.6 \times 10^{-4}$. *Dotted line: The output is limited to 1 with* $\hat{\rho}_{1NL} = 1 \times 10^5$ *and* $\hat{\rho}_{2NL} = -1 \times 10^5$. *Dashed line: The limited level is 0.81 with* $\hat{\rho}_{1NL} = 1.1 \times 10^5$ *and* $\hat{\rho}_{2NL} = -1.1 \times 10^5$. *Solid line: The limited level is 0.64 with* $\hat{\rho}_{1NL} = 1.25 \times 10^5$ *and* $\hat{\rho}_{2NL} = -1.25 \times 10^5$ *[14].*

2.5 Future challenge

Metamaterials are required to realize the negative refractive index. These artificial materials were introduced in 1968 for electromagnetic waves [19]. More efforts were made to implement these materials afterward [20–27]. Furthermore, the nonlinear properties of metamaterials are also of interest [28–30]. Phononic metamaterials can also be designed so that their effective Young's modulus or mass density is negative [31–33]. Realization of the phononic hard limiter requires the nonlinear metamaterial. Research in this area can be an exciting subject and help design a variety of devices in phononics. Furthermore, a phononic limiter can be utilized to manipulate phonon-photon interaction in optomechanics and electron-phonon interaction in optoelectronic or electronic devices.

3. Phononic wave coupling in slab waveguides

Understanding the wave coupling is crucial to design devices based on wave propagation. The Coupled mode theory (CMT) discusses the energy exchange between different modes or coupled waveguides [34–36]. CMT in the base of Machzender interferometer, microsphere resonators, couplers, switches, and filters in photonics [37]. Herein, the CMT is applied to the elastic slab waveguide to obtain the coupled mode equations (CME) for shear and Rayleigh-Lame modes inspired by electromagnetic wave propagation.

3.1 Coupled mode equations

3.1.1 Shear waves

The wave equation for the shear waves (SH) is expressed by Eq. (37) for a waveguide orientated in the direction of z (**Figure 5**), where $\varepsilon^0 = \rho/_{\mu}$. This equation is for an unperturbed system containing a single waveguide. The solution for the shear wave equation is given by Eq. (42), in which a_m, $U_m(x)$, and β_m are the amplitude, mode shape, and wave number, respectively, and m is the mode number.

$$\left(\frac{\partial^2}{\partial x^2}+\frac{\partial^2}{\partial z^2}\right)u_y+\varepsilon^0\omega^2 u_y=0 \tag{41}$$

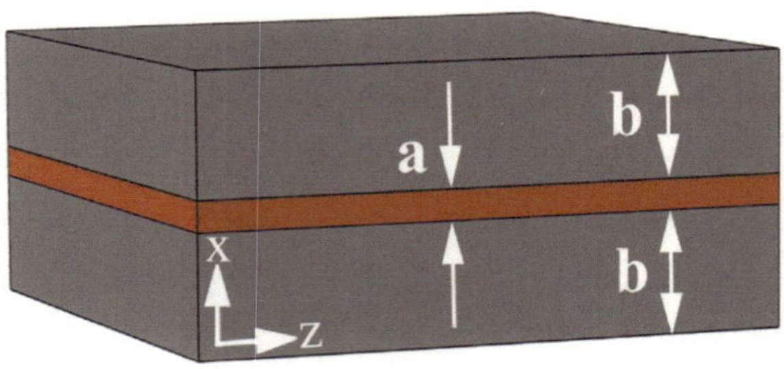

Figure 5.
The elastic slab waveguide. a and b are the width of the core and cladding regions [38].

$$u_y = \sum_m a_{ym} U_{ym}(x) e^{-j\beta_m z} \tag{42}$$

The wave equation for the mode shape is obtained via inserting Eq. (42) in Eq. (41) and given by:

$$\frac{\partial^2 U_{ym}(x)}{\partial x^2} + \left(\varepsilon^0 \omega^2 - \beta_m^2\right) U_{ym}(x) = 0 \tag{43}$$

For the waveguides that interchange energy, assuming they are not very close together, the perturbed wave equation is written as Eq. (44). ε^1 represents the perturbation.

$$\left(\frac{\partial^2}{\partial x^2} + \frac{\partial^2}{\partial z^2}\right) u_y + \left(\varepsilon^0 + \varepsilon^1\right)\omega^2 u_y = 0 \tag{44}$$

The perturbation causes the amplitude not to be constant in the propagation direction; hence, the displacement should be written as Eq. (45), where $a_{ym}(z)$ is the slowly varying amplitude.

$$u_y = \sum_m a_{ym}(z) U_{ym}(x) e^{-j\beta_m z} \tag{45}$$

Inserting Eq. (45) in Eq. (44) results in Eq. (46).

$$\begin{aligned} &\sum_m a_{ym}(z)\left(\frac{\partial^2 U_{ym}(x)}{\partial x^2} + \left(\varepsilon^0\omega^2 - \beta_m^2\right) U_{ym}(x)\right) e^{-j\beta_m z} + \\ &\sum_m \left(\frac{\partial^2 a_{ym}(z)}{\partial z^2} - j2\beta_m \frac{\partial a_{ym}(z)}{\partial z}\right) U_{ym}(x) e^{-j\beta_m z} \\ &= -\varepsilon^1 \omega^2 \sum_m a_{ym}(z) U_{ym}(x) e^{-j\beta_m z} \end{aligned} \tag{46}$$

Eq. (43) specifies that the first term in Eq. (46) is zero. Furthermore, the second derivative of amplitude is neglected because of its slow variation. The orthogonality of modes leads to Eq. (47) with the coupling coefficient obtained by Eq. (48) [38].

$$\frac{\partial a_{yn}(z)}{\partial z} = -j\frac{\omega^2}{2\beta_n} \sum_m \kappa_{(SH)n,m} a_{ym}(z) e^{-j(\beta_m - \beta_n)z} \tag{47}$$

$$\kappa_{(SH)n,m} = \frac{\int_{-\infty}^{+\infty} U_{yn}^*(x) \varepsilon^1 U_{ym}(x) dx}{\int_{-\infty}^{+\infty} U_{yn}^*(x) U_{yn}(x) dx} \tag{48}$$

3.1.2 Rayleigh-lame waves

Rayleigh-Lame (RL) modes have displacement in the x and z directions for the waveguide depicted in **Figure 5**. Therefore, the wave equation for the RL modes is expressed by [15]:

$$\mu \nabla^2 u_i + (\lambda + \mu) \frac{\partial}{\partial x_i} \sum_{i=1,2} \frac{\partial u_i}{\partial x_i} + \rho \omega^2 u_i = 0 \tag{49}$$

where i = 1 and 2 indicate the direction of x and z, respectively. Consequently, the wave equations for displacement components are given by:

$$\eta_s^0 \left(\frac{\partial^2 u_x}{\partial x^2} + \frac{\partial^2 u_x}{\partial z^2} \right) + \eta_l^0 \left(\frac{\partial^2 u_x}{\partial x^2} + \frac{\partial^2 u_z}{\partial x \partial z} \right) + \omega^2 u_x = 0 \tag{50}$$

$$\eta_s^0 \left(\frac{\partial^2 u_z}{\partial x^2} + \frac{\partial^2 u_z}{\partial z^2} \right) + \eta_l^0 \left(\frac{\partial^2 u_z}{\partial x^2} + \frac{\partial^2 u_x}{\partial x \partial z} \right) + \omega^2 u_z = 0 \tag{51}$$

where $\eta_s^0 = \mu/\rho$ and $\eta_l^0 = (\lambda + \mu)/\rho$ are unperturbed parameters of the waveguide. Coupling results in the perturbation in η_s^0 and η_l^0; hence, the above equations change to Eqs. (52) and (53), where η_s^1 and η_l^1 are the perturbation terms.

$$\left(\eta_s^0 + \eta_s^1\right) \left(\frac{\partial^2 u_x}{\partial x^2} + \frac{\partial^2 u_x}{\partial z^2} \right) + \left(\eta_l^0 + \eta_l^1\right) \left(\frac{\partial^2 u_x}{\partial x^2} + \frac{\partial^2 u_z}{\partial x \partial z} \right) + \omega^2 u_x = 0 \tag{52}$$

$$\left(\eta_s^0 + \eta_s^1\right) \left(\frac{\partial^2 u_z}{\partial x^2} + \frac{\partial^2 u_z}{\partial z^2} \right) + \left(\eta_l^0 + \eta_l^1\right) \left(\frac{\partial^2 u_z}{\partial x^2} + \frac{\partial^2 u_x}{\partial x \partial z} \right) + \omega^2 u_z = 0 \tag{53}$$

Similar to the shear waves, the displacement components (u_x and u_z) are assumed to be:

$$u_x = \sum_m a_{xm}(z) U_{xm}(x) e^{-j\beta_m z} \tag{54}$$

$$u_x = \sum_m a_{xm}(z) U_{xm}(x) e^{-j\beta_m z} \tag{55}$$

CMEs for RL modes are expressed by Eqs. (56) and (57) using the solutions mentioned above, the orthogonality of modes, and the slow variation of amplitude. Eqs. (58) and (59) give the coupling coefficients in CMEs [38].

$$2j\beta_n \frac{\partial a_{xn}}{\partial z} + \beta_n^2 a_{xn} = \sum_m \left[\kappa_{(xx)n,m} a_{xm} + \kappa'_{(xz)n,m} \left(\frac{\partial a_{zm}}{\partial z} - j\beta_m a_{zm} \right) \right] e^{-j(\beta_m - \beta_n)z} \tag{56}$$

$$2j\beta_n \frac{\partial a_{zn}}{\partial z} + \beta_n^2 a_{zn} = \sum_m \left[\kappa_{(zz)n,m} a_{zm} + \kappa'_{(zx)n,m} \left(\frac{\partial a_{xm}}{\partial z} - j\beta_m a_{xm} \right) \right] e^{-j(\beta_m - \beta_n)z} \tag{57}$$

$$\kappa_{(xx)n,m} = \frac{\int_{-\infty}^{+\infty} U_{xn}^*(x) \left(1 + \frac{\eta_l^1(x)}{\eta_s^1(x)} \right) \frac{\partial^2 U_{xm}(x)}{\partial x^2} dx}{\int_{-\infty}^{+\infty} U_{xn}^*(x) U_{xn}(x) dx} \tag{58}$$

$$\kappa'_{(xz)n,m} = \frac{\int_{-\infty}^{+\infty} U^*_{xn}(x)\left(\frac{\eta^1_l(x)}{\eta^1_s(x)}\right)\frac{\partial U_{zm}(x)}{\partial x}dx}{\int_{-\infty}^{+\infty} U^*_{xn}(x)U_{xn}(x)dx} \tag{59}$$

$$\kappa_{(zz)n,m} = \frac{\int_{-\infty}^{+\infty} U^*_{zn}(x)\left(1+\frac{\eta^1_l(x)}{\eta^1_s(x)}\right)\frac{\partial^2 U_{zm}(x)}{\partial x^2}dx}{\int_{-\infty}^{+\infty} U^*_{zn}(x)U_{zn}(x)dx} \tag{60}$$

$$\kappa'_{(zx)n,m} = \frac{\int_{-\infty}^{+\infty} U^*_{zn}(x)\left(\frac{\eta^1_l(x)}{\eta^1_s(x)}\right)\frac{\partial U_{xm}(x)}{\partial x}dx}{\int_{-\infty}^{+\infty} U^*_{zn}(x)U_{zn}(x)dx} \tag{61}$$

The modal analysis should be performed to obtain the mode shapes for calculating the coupling coefficients. Eqs. (56) and (57) can be simplified by assuming $\hat{a}_{xn} \triangleq a_{xn}e^{-j\beta_n z/2}$, $\hat{a}_{zn} \triangleq a_{zn}e^{-j\beta_n z/2}$, $\hat{a}_{xm} \triangleq a_{xm}e^{-j\beta_m z/2}$, and $\hat{a}_{zm} \triangleq a_{zm}e^{-j\beta_m z/2}$ given by:

$$\frac{\partial \hat{a}_{xn}}{\partial z} = -\frac{j}{2\beta_n}\sum_m\left[\kappa_{(xx)n,m}\hat{a}_{xm} + \kappa'_{(xz)n,m}\left(\frac{\partial \hat{a}_{zm}}{\partial z} - j\frac{\beta_m}{2}\hat{a}_{zm}\right)\right]e^{-\frac{j(\beta_m-\beta_n)z}{2}} \tag{62}$$

$$\frac{\partial \hat{a}_{zn}}{\partial z} = -\frac{j}{2\beta_n}\sum_m\left[\kappa_{(zz)n,m}\hat{a}_{zm} + \kappa'_{(zx)n,m}\left(\frac{\partial \hat{a}_{xm}}{\partial z} - j\frac{\beta_m}{2}\hat{a}_{xm}\right)\right]e^{-\frac{j(\beta_m-\beta_n)z}{2}} \tag{63}$$

3.2 Two coupled waveguides

3.2.1 SH modes

The couple mode equations for two single-mode waveguides, illustrated in **Figure 6**, are given by Eqs. (64) and (65), using Eq. (47). The indices 1 and 2 refer to the two coupled waveguides.

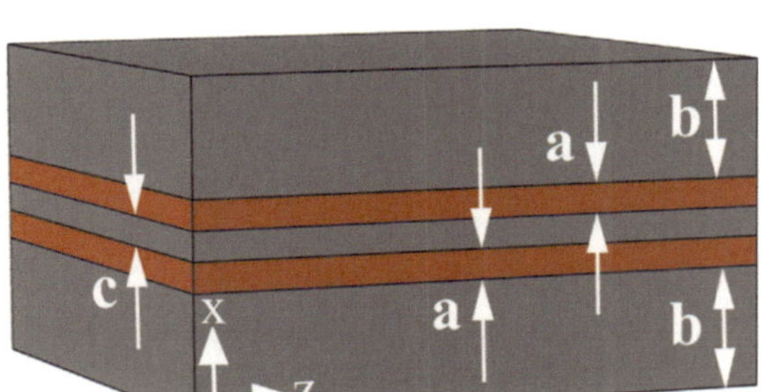

Figure 6.
Two coupled waveguides. c represents the distance between waveguides. a and b are the widths of the core and cladding regions [38].

$$\frac{\partial a_{y1}(z)}{\partial z} = -j\frac{\omega^2}{2\beta_1}\kappa_{(SH)1,2}a_{y2}(z)e^{-j(\beta_2-\beta_1)z} \tag{64}$$

$$\frac{\partial a_{y2}(z)}{\partial z} = -j\frac{\omega^2}{2\beta_2}\kappa_{(SH)2,1}a_{y1}(z)e^{-j(\beta_1-\beta_2)z} \tag{65}$$

The CMEs are solved for two coupled waveguides, shown in **Figure 6**, considering both core width (a) and distance between waveguides (c) to be 200 μm and the cladding width (b) to be 4 mm to evaluate the coupling process. The result is illustrated in **Figure 7**. The materials of core and cladding regions are presumed copper and silicon, respectively. The coupling gives rise to the energy transfer between waveguides periodically. The value of elastic energy in each waveguide depends on the length of the coupled system [38].

3.2.2 RL modes

The CMEs of Rayleigh-Lame modes for the coupled waveguides in **Figure 6** are given by Eqs. (66–69) using Eqs. (62) and (63).

$$\frac{\partial \hat{a}_{x1}}{\partial z} = -\frac{j}{2\beta_1}\left[\kappa_{(xx)1,2}\hat{a}_{x2} + \kappa'_{(xz)1,2}\left(\frac{\partial \hat{a}_{z2}}{\partial z} - j\frac{\beta_2}{2}\hat{a}_{z2}\right)\right]e^{-\frac{j(\beta_2-\beta_1)z}{2}} \tag{66}$$

$$\frac{\partial \hat{a}_{z1}}{\partial z} = -\frac{j}{2\beta_1}\left[\kappa_{(zz)1,2}\hat{a}_{z2} + \kappa'_{(zx)1,2}\left(\frac{\partial \hat{a}_{x2}}{\partial z} - j\frac{\beta_2}{2}\hat{a}_{x2}\right)\right]e^{-\frac{j(\beta_2-\beta_1)z}{2}} \tag{67}$$

$$\frac{\partial \hat{a}_{x2}}{\partial z} = -\frac{j}{2\beta_2}\left[\kappa_{(xx)2,1}\hat{a}_{x1} + \kappa'_{(xz)2,1}\left(\frac{\partial \hat{a}_{z1}}{\partial z} - j\frac{\beta_1}{2}\hat{a}_{z1}\right)\right]e^{-\frac{j(\beta_1-\beta_2)z}{2}} \tag{68}$$

$$\frac{\partial \hat{a}_{z2}}{\partial z} = -\frac{j}{2\beta_1}\left[\kappa_{(zz)2,1}\hat{a}_{z1} + \kappa'_{(zx)2,1}\left(\frac{\partial \hat{a}_{x1}}{\partial z} - j\frac{\beta_1}{2}\hat{a}_{x1}\right)\right]e^{-\frac{j(\beta_1-\beta_2)z}{2}} \tag{69}$$

The displacement components in one waveguide are coupled to both displacement components of another one. The CMEs for the waveguide are solved with the dimensions mentioned in the previous section. The modal analysis should be first performed to calculate the coupling coefficients. **Figure 8** illustrates the coupling results when the input displacement is considered in the x direction. **Figure 8a** and **b** depict the amplitudes in the x and z directions, respectively. On the contrary, in **Figure 9**, the input displacement is totally in the z-direction. **Figure 9a** and **b** are for displacement amplitude in the x and z directions, respectively.

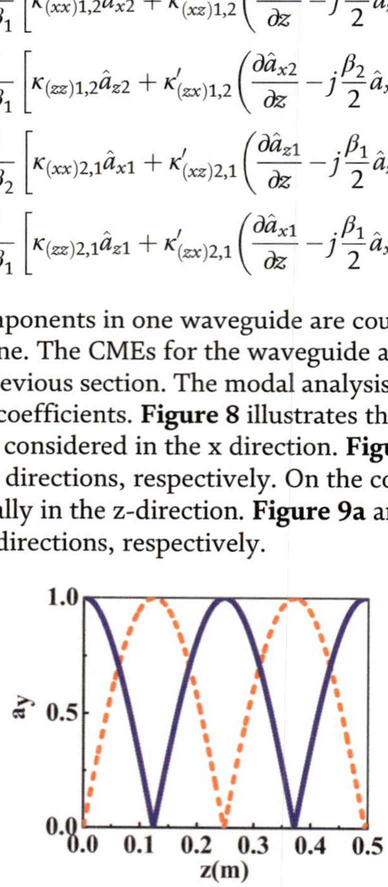

Figure 7.
The amplitudes of waves in two coupled single-mode waveguides for the coupling of SH modes [38].

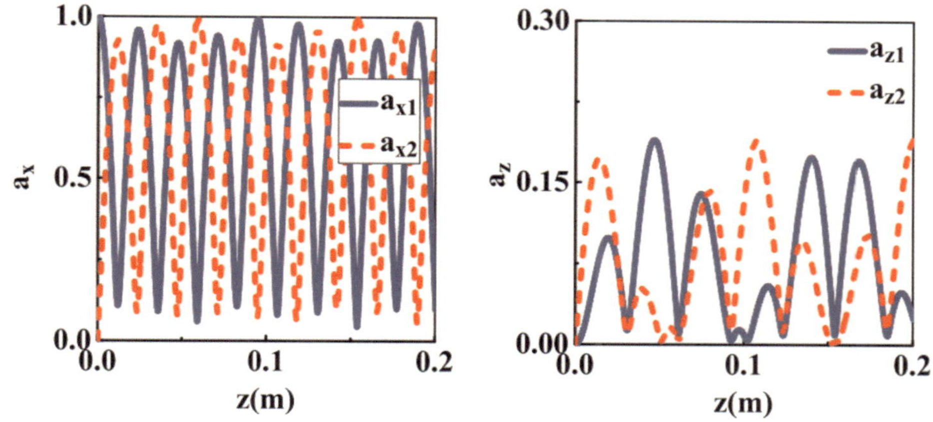

(a) (b)

Figure 8.
The coupling process for the first RL mode when $\hat{a}_{x1}(0) = 1$, $\hat{a}_{z1}(0) = 0$, $\hat{a}_{x2}(0) = 0$, and $\hat{a}_{z2}(0) = 0$. (a) and (b) are for amplitudes of displacements in the x and z directions, respectively [38].

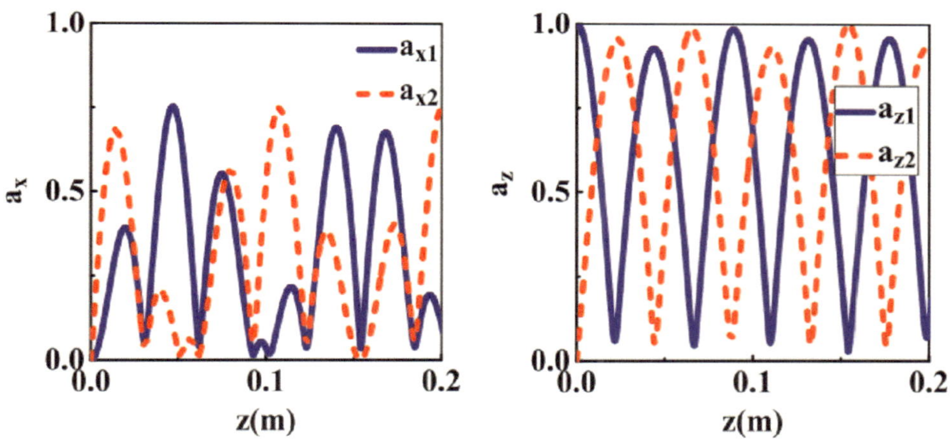

(a) *(b)*

Figure 9.
The coupling process for the first RL mode when $\hat{a}_{x1}(0) = 0$, $\hat{a}_{z1}(0) = 1$, $\hat{a}_{x2}(0) = 0$, and $\hat{a}_{z2}(0) = 0$. (a) and (b) are for amplitudes of displacements in the x and z directions, respectively [38].

3.2.3 SH wave coupling between non-parallel waveguides

The SH modes of a slab waveguide are given by Eq. (70), where t and C are the thickness of the core and normalization constant, respectively. The indices co and cl indicate the core and cladding regions. p and h are defined as Eqs. (71) and (72).

$$U_y(x) = \begin{cases} Ce^{-px}, & 0 \le x < +\infty \\ C\left[\cos(hx) - \dfrac{\mu_{cl}p}{\mu_{co}h}\sin(hx)\right], & -t \le x \le 0 \\ C\left[\cos(ht) + \dfrac{\mu_{cl}p}{\mu_{co}h}\sin(ht)\right]e^{p(x+t)}, & -\infty \le x < -t \end{cases} \tag{70}$$

$$p = \left(\beta^2 - n_{cl}^2\omega^2\right)^{½} \tag{71}$$

$$h = \left(n_{co}^2\omega^2 - \beta^2\right)^{½} \tag{72}$$

n_{co} and n_{cl} are the refractive indices of the core and cladding, equaling the square root of ε^0. The elastic power flux is expressed as Eq. (73), in which σ is the stress tensor [39].

$$\iint -\left(\sigma.\frac{\partial \vec{u}}{\partial t}\right).\hat{n}dS \tag{73}$$

The normalization constant is calculated by Eq. (74), assuming the power flux to be 1.

$$|C| = \frac{(\mu_{co}h)^2 + (\mu_{cl}p)^2}{\mu_{co}\sqrt{2\beta\omega(\mu_{cl}p)^3}} \tag{74}$$

The above-introduced modes are applied to the structure containing two non-parallel coupled waveguides illustrated in **Figure 10** to investigate the coupling process. The displacement for SH waves can be written as Eq. (75) when the waveguides are not very close. The wave equation for the total displacement is given by Eq. (76), where $\Delta n_{co1}^2(x_1) = n_{co1}^2(x_1) - n_{cl}^2$ and $\Delta n_{co2}^2(x_2) = n_{co2}^2(x_2) - n_{cl}^2$ [40].

$$u_y = a_{y1}(z_1)U_{y1}(x_1)e^{-j\beta z_1} + a_{y2}(z_2)U_{y2}(x_2)e^{-j\beta z_2} \tag{75}$$

$$\left(\frac{\partial^2}{\partial x^2} + \frac{\partial^2}{\partial z^2}\right)u_y + \omega^2\left[n_{cl}^2 + \Delta n_{co1}^2(x_1) + \Delta n_{co2}^2(x_2)\right]u_y = 0 \tag{76}$$

Eqs. (79–82) gives the CMEs for the structure depicted in **Figure 10**, considering the orthogonality of modes and slowly varying amplitude approximation. The coupling coefficients are calculated through Eqs. (79–82) [38].

$$\frac{\partial a_{y1}(z_1)}{\partial z_1} = -j\frac{\omega^2}{2\beta}\left[\kappa_{12}a_{y2}(z_2)e^{j\beta(1-\cos\alpha)} + \kappa_{11}a_{y1}(z_1)\right] \tag{77}$$

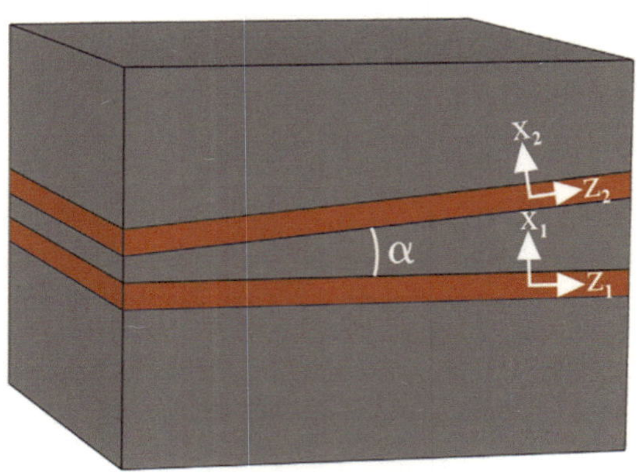

Figure 10.
Two non-parallel coupled waveguides [38].

$$\frac{\partial a_{y2}(z_2)}{\partial z_2} = -j\frac{\omega^2}{2\beta}\left[\kappa_{21}a_{y1}(z_1)e^{j\beta(1-\cos\alpha)} + \kappa_{22}a_{y2}(z_2)\right] \tag{78}$$

$$\kappa_{11} = \frac{\int_{-\infty}^{+\infty} U_{y1}^*(x_1)\Delta n_{co2}^2(x_1,z_1)U_{y1}(x_1)dx_1}{\int_{-\infty}^{+\infty} U_{y1}^*(x_1)U_{y1}(x_1)dx_1} \tag{79}$$

$$\kappa_{12} = \frac{\int_{-\infty}^{+\infty} U_{y1}^*(x_1)\Delta n_{co1}^2(x_1)U_{y2}(x_1,z_1)e^{-j\beta\sin\alpha x_1}dx_1}{\int_{-\infty}^{+\infty} U_{y1}^*(x_1)U_{y1}(x_1)dx_1} \tag{80}$$

$$\kappa_{22} = \frac{\int_{-\infty}^{+\infty} U_{y2}^*(x_2)\Delta n_{co1}^2(x_2,z_2)U_{y2}(x_2)dx_2}{\int_{-\infty}^{+\infty} U_{y2}^*(x_2)U_{y2}(x_2)dx_2} \tag{81}$$

$$\kappa_{21} = \frac{\int_{-\infty}^{+\infty} U_{y2}^*(x_2)\Delta n_{co2}^2(x_2)U_{y1}(x_2,z_2)e^{-j\beta\sin\alpha x_2}dx_2}{\int_{-\infty}^{+\infty} U_{y2}^*(x_2)U_{y2}(x_2)dx_2} \tag{82}$$

For small angles between waveguides, the coupling coefficients are obtained by [38]:

$$\kappa_{11} = \frac{(h^2+p^2)(e^{2pt}-1)}{\omega^2\left(2+\frac{2\mu_{cl}p^2}{\mu_{co}h^2}+\left(1+\frac{\mu_{cl}^2p^2}{\mu_{co}^2h^2}\right)pt\right)}e^{-2pz_2\sin\alpha} = \kappa_0 e^{-2pz_1\tan\alpha} \tag{83}$$

$$\kappa_{12} = \frac{2p^2\left(\left(1+\frac{\mu_{cl}}{\mu_{co}}\right)(e^{pt}+1)-2\right)}{\omega^2\left(2+\frac{2\mu_{cl}p^2}{\mu_{co}h^2}+\left(1+\frac{\mu_{cl}^2p^2}{\mu_{co}^2h^2}\right)pt\right)}e^{-pz_1\sin\alpha} = \kappa_1 e^{-pz_1\sin\alpha} \tag{84}$$

$$\kappa_{22} = \frac{(h^2+p^2)(e^{2pt}-1)}{\omega^2\left(2+\frac{2\mu_{cl}p^2}{\mu_{co}h^2}+\left(1+\frac{\mu_{cl}^2p^2}{\mu_{co}^2h^2}\right)pt\right)}e^{-2pz_1\sin\alpha} = \kappa_0 e^{-2pz_2\tan\alpha} \tag{85}$$

$$\kappa_{21} = \frac{2p^2\left(\left(1+\frac{\mu_{cl}}{\mu_{co}}\right)(e^{pt}+1)-2\right)}{\omega^2\left(2+\frac{2\mu_{cl}p^2}{\mu_{co}h^2}+\left(1+\frac{\mu_{cl}^2p^2}{\mu_{co}^2h^2}\right)pt\right)}e^{-pz_2\sin\alpha} = \kappa_1 e^{-pz_2\sin\alpha} \tag{86}$$

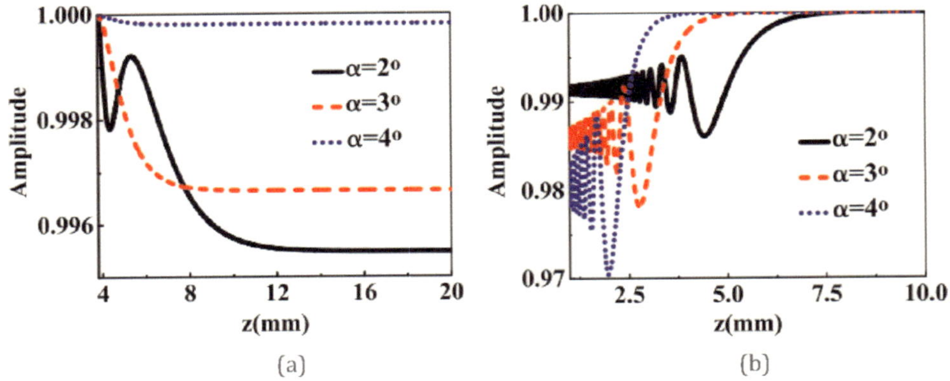

Figure 11.
(a) Two waveguides are moving away from each other. (b) Two waveguides are getting closer to each other [38].

The CMEs for the waveguides, whose core and cladding are similar to section 3.2.1, are solved, and the results are illustrated in **Figure 11** for two different situations: the waveguides getting away from each other and vice versa. In the case that two waveguides move away from each other after a certain length due to the increase in the distance between the two waveguides, the interaction between them tends to zero, and the coupling stops. Consequently, the amplitude remains constant. Conversely, when two waveguides gradually get closer to each other, the intensity of coupling increases, and as a result, the elastic power is exchanged faster between two waveguides.

3.3 Future challenges

Modal analysis of waveguides with inhomogeneity, such as impurities in materials or rough boundaries between the core and cladding regions, which cause mode coupling, can help to reach a deep understanding of the wave propagation in waveguides. Furthermore, investigation of the coupling in a system including more than two waveguides may be required to know how we can guide the desired value of elastic energy toward a specific point in optoelectronic or photonic devices. Moreover, exploring the quantum mechanical behavior of phonons, named by quantum acoustics, is a significant issue because of the growing demand for quantum computing; hence, designing devices that can generate, transport, and detect a single phonon is required.

4. Conclusion

Since phonons have an essential role in physics, designing suitable devices that allow physicists and engineers to control and manipulate them and their interaction with electrons and photons has been of interest to researchers. As nonlinearity is widely used to design desirable devices in electromagnetics, inspired by photonics and the similarity between photons and phonons, the nonlinear Schrödinger equation (NSE), which describes the nonlinear wave propagation in nonlinear and dispersive media, was developed for phononic waves. A phononic wave hard limiter, which confines the amplitude of the output wave to a desirable value, was introduced using

NSE. The coupled mode theory was also developed for elastic slab waveguides, and the coupled mode equations are obtained for shear and Rayleigh-Lame modes. Finally, the shear wave coupling was investigated analytically for two non-parallel slab waveguides.

Author details

Ali Rostami[1*] and Hodjat Ahmadi[2]

1 University of Tabriz, Tabriz, Islamic Republic of Iran

2 ASEPE Company, Tabriz, Islamic Republic of Iran

*Address all correspondence to: rostami@tabrizu.ac.ir

References

[1] Delsing P et al. The 2019 surface acoustic waves roadmap. Journal of Physics D: Applied Physics. 2019;**52**(35): 353001

[2] Giustino F. Electron-phonon interactions from first principles. Reviews of Modern Physics. 2017;**89**(1): 015003

[3] Efetov DK, Kim P. Controlling electron-phonon interactions in graphene at ultrahigh carrier densities. Physical Review Letters. 2010;**105**(25): 256805

[4] Safavi-Naeini AH et al. Electromagnetically induced transparency and slow light with optomechanics. Nature. 2011; **472**(7341):69

[5] Chang D, Safavi-Naeini AH, Hafezi M, Painter O. Slowing and stopping light using an optomechanical crystal array. New Journal of Physics. 2011;**13**(2):023003

[6] Bell DA. Solid State Pulse Circuits. Reston, VA: Reston Publishing Co., Inc; 1981 474 p

[7] Filanovsky I, Baltes H. CMOS Schmitt trigger design. IEEE Transactions on Circuits and Systems I: Fundamental Theory and Applications. 1994;**41**(1): 46-49

[8] Brzozowski L, Sargent EHT. All-optical analog-to-digital converters, hard limiters, and logic gates. Journal of Lightwave Technology. 2001;**19**(1):114

[9] Rostami A, Rostami G. Full optical analog to digital (A/D) converter based on Kerr-like nonlinear ring resonator. Optics Communications. 2003;**228**(1): 39-48

[10] Streets AM, Li A, Chen T, Huang Y. Imaging without Fluorescence: Nonlinear Optical Microscopy for Quantitative Cellular Imaging. USA: ACS Publications; 2014

[11] Agrawal GP. Nonlinear Fiber Optics. USA: Academic Press; 2007

[12] McCall K. Theoretical study of nonlinear elastic wave propagation. Journal of Geophysical Research: Solid Earth. 1994;**99**(B2):2591-2600

[13] Rose JL. Ultrasonic Guided Waves in Solid Media. UK: Cambridge University Press; 2014

[14] Ahmadi H, Rostami A. Phononic wave hard limiter. Journal of Sound and Vibration. 2019;**443**:230-237

[15] Morse R. The velocity of compressional waves in rods of rectangular cross-section. The Journal of the Acoustical Society of America. 1950; **22**(2):219-223

[16] Waldron R. Some problems in the theory of guided microsonic waves. IEEE Transactions on Microwave Theory and Techniques. 1969;**17**(11):893-904

[17] Ding Y, Liu Z, Qiu C, Shi J. Metamaterial with simultaneously negative bulk modulus and mass density. Physical Review Letters. 2007;**99**(9): 093904

[18] He J, Cada M. Optical bistability in semiconductor periodic structures. IEEE Journal of Quantum Electronics. 1991; **27**(5):1182-1188

[19] Veselago VG. The electrodynamics of substances with simultaneously negative values of and μ. Soviet Physics Uspekhi. 1968;**10**(4):509

[20] Shelby RA, Smith DR, Schultz S. Experimental verification of a negative index of refraction. Science. 2001; **292**(5514):77-79

[21] Ling F, Zhong Z, Huang R, Zhang B. A broadband tunable terahertz negative refractive index metamaterial. Scientific Reports. 2018;**8**(1):9843

[22] Suzuki T, Sekiya M, Sato T, Takebayashi Y. Negative refractive index metamaterial with high transmission, low reflection, and low loss in the terahertz waveband. Optics Express. 2018;**26**(7):8314-8324

[23] Nguyen HT et al. Broadband negative refractive index obtained by plasmonic hybridization in metamaterials. Applied Physics Letters. 2016;**109**(22):221902

[24] Ling F, Zhong Z, Zhang Y, Huang R, Zhang B. Broadband negative-refractive index terahertz metamaterial with optically tunable equivalent-energy level. Optics Express. 2018;**26**(23): 30085-30099

[25] Yoo S, Lee S, Park Q-H. Loss-free negative-index metamaterials using forward light scattering in dielectric meta-atoms. ACS Photonics. 2018;**5**(4): 1370-1374

[26] Lee G-H, Park G-H, Lee H-J. Observation of negative refraction of Dirac fermions in graphene. Nature Physics. 2015;**11**(11):925

[27] Liang Y, Yu Z, Ruan N, Sun Q, Xu T. Freestanding optical negative-index metamaterials of green light. Optics Letters. 2017;**42**(16):3239-3242

[28] Lapine M, Shadrivov IV, Kivshar YS. Colloquium: nonlinear metamaterials. Reviews of Modern Physics. 2014;**86**(3): 1093

[29] Shadrivov IV, Kozyrev AB, van der Weide DW, Kivshar YS. Tunable transmission and harmonic generation in nonlinear metamaterials. Applied Physics Letters. 2008;**93**(16):161903

[30] Liu Y, Bartal G, Genov DA, Zhang X. Subwavelength discrete solitons in nonlinear metamaterials. Physical Review Letters. 2007;**99**(15):153901

[31] Liu Z et al. Locally resonant sonic materials. Science. 2000;**289**(5485): 1734-1736

[32] Liu Z, Chan CT, Sheng P. Analytic model of phononic crystals with local resonances. Physical Review B. 2005; **71**(1):014103

[33] Fang N et al. Ultrasonic metamaterials with negative modulus. Nature Materials. 2006;**5**(6):452

[34] Yariv A. Coupled-mode theory for guided-wave optics. IEEE Journal of Quantum Electronics. 1973;**9**(9):919-933

[35] Nishihara H, Haruna M, Suhara T. Optical Integrated Circuits. New York: McGraw-Hill; 1989

[36] Okamoto K. Fundamentals of Optical Waveguides. UK: Academic Press; 2006

[37] Pollock CR, Lipson M. Integrated Photonics (no. 25). USA: Springer; 2003

[38] Hashemzadeh P, Ahmadi H, Rostami A. Generalized coupled mode theory in phononic slab waveguides. Optik. 2022;**249**:168222

[39] Snieder R. General theory of elastic wave scattering. In: Scattering. USA: Elsevier; 2002. pp. 528-542

[40] Liang H, Shi S, Ma L. Coupled-mode theory of nonparallel optical waveguides. Journal of Lightwave Technology. 2007; **25**(8):2233-2235

Chapter 6

Band Gap Calculation Theory of Periodic Structure and Its Application in Vibration Reduction of Track Structure

Wenjie Guo

Abstract

As the construction of the railway network becomes more and more mature, how to operate and maintain it has become a problem that cannot be ignored. Among them, the vibration and noise problems in train operation have become a hot issue in rail transit operation and maintenance. Although the railway track structure has various forms, its laying method has obvious periodicity, so the phononic crystal theory can be used to study the propagation characteristics of elastic waves in the track structure. Based on this, this chapter introduces the phononic crystal theory into the track structure with a regular period, calculates the band gap characteristics under various track structure forms, and studies the propagation law of elastic wave in it, so as to guide the vibration and noise reduction control in the field of rail transit.

Keywords: band gap calculation, periodic structure, vibration reduction, track structure, phononic crystal theory

1. Introduction

The track structure is composed of rail, sleeper, turnout, connecting parts, anti-climbing equipment, ballast bed, roadbed, and so on. This chapter will focus on the CRTS-III track and the CRTS-II track. We can assume that the orbital structure is periodically arranged, and when the elastic wave is transmitted in the periodic structure [1–4], there is a unique filtering property that enables the wave of certain specific frequencies to be transmitted in the structure. At this time, the wave of other frequencies will be attenuated. This unique physical phenomenon is the band gap characteristic. In this section, the phononic crystal theory [5, 6] is employed to confirm the characteristics of elastic wave propagation band gaps within the orbital structure. Starting from the theoretical model, the two beam-plate coupling models of continuous beam-plate model [7] and discrete beam-plate model are introduced. The characteristics of elastic wave propagation band gaps in the CRTS-III plate and CRTS-II plate are calculated by combining the model with the actual structure [8–11].

2. Theoretical model

2.1 Two types of beam-plate coupling models for plates

Two types of track structure are simplified as an infinite structure composed of rail with the periodic nature [12, 13], fastener, track slab, and CA layer (or SCC layer). The fastener and CA layer are simplified as spring elements. To take the bending and shear properties of the track slab and rail into consideration at the same time, the rail is assumed to be a Timoshenko beam element, and the track slab is simplified as a Mindlin plate element. Then the CRTS-III plate and CRTS-II plate are simplified into two beam-plate coupling models with one period. **Figure 1** (a) illustrates the discrete beam-plate model, whereas (b) depicts the continuous beam-plate model. The plates are arranged longitudinally along the beam and are periodic. **Figure 2** shows the structural units in a unit cell. The model shown in **Figure 2** satisfies the conditions beneath: (1) The beam and the plate are linked by the periodic distribution of line

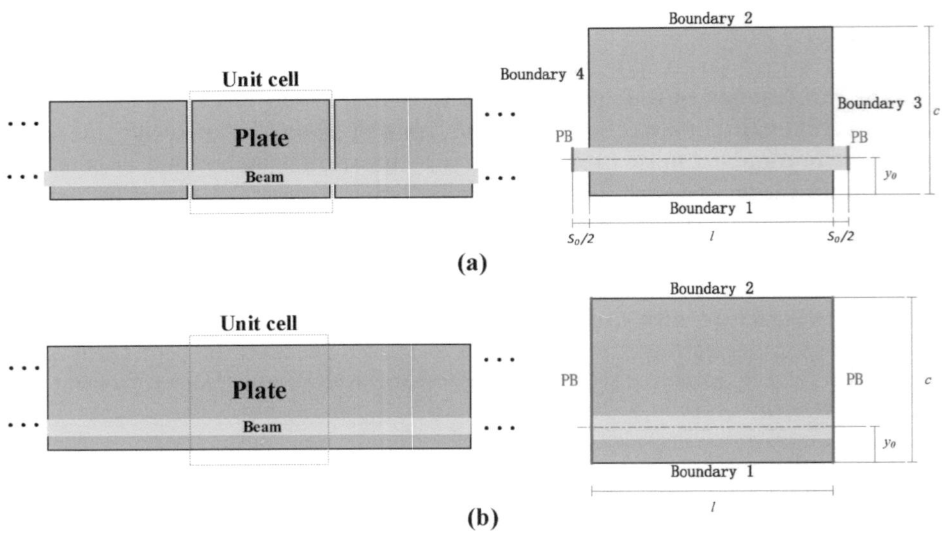

Figure 1.
The unit cell of two beam-plate models (Figure (a) is the model of CRTS-III plate; Figure (b) is the model of CRTS-II plate).

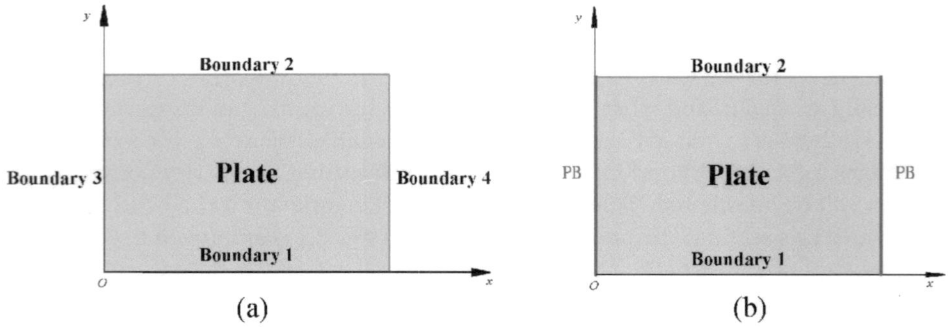

Figure 2.
The plate boundary of the model.

springs; (2) In the first model, the adjacent plates are separated by a gap. B2 is the symmetric boundary, B1, B3, and B4 are the free boundary; (3) Boundary 3 and boundary 4 have a periodic nature, and no gap exists between two adjacent plates in the second model.

2.2 Theoretical expression

In this chapter, the total system energy consists of many aspects, which include: (1) The energy components of the rail, including both strain and kinetic energies; (2) The rail's energy profile encompasses both strain and kinetic energy; (3) The potential energy within the CA layer situated beneath the plate; (4) The fasteners at the beam-plate joint possess elastic potential energy. Meanwhile, this chapter only considers the vertical vibration of the beam-slab coupling model.

2.2.1 The functional form of the beam's energy

A rail is taken for analysis. To completely take the result into consideration, which is the influence of two modes of rail bending and shear on band gap characteristics and vertical vibration, we use the Timoshenko beam model to describe the rail. The model aims to characterize the rail's dynamic response. The vibration differential equation is [14]:

$$\begin{cases} \dfrac{d}{dx'}\left[\kappa_B G_B A_B\left(\dfrac{dv}{dx'}-\gamma\right)\right]+\rho_B A_B \omega^2 v=0, \\ \dfrac{d}{dx'}\left(E_B I_B \dfrac{d\gamma}{dx'}\right)+\kappa_B G_B A_B\left(\dfrac{dv}{dx'}-\gamma\right)+\rho_B I_B \omega^2 \gamma=0, \end{cases} \tag{1}$$

where v is vertical displacement; γ is the rail rotation angle; ρ_B is rail density; Young's modulus of the rail is represented by E_B; The moment of inertia of the cross-section is denoted by I_B; the shear coefficient is denoted by κ_B; κ_B is shear modulus and equal to $E_B/[2(1+\mu_B)]$; the Poisson's ratio is denoted by μ_B; the cross-sectional area is denoted by A_B; ω is the structural characteristic frequency.

Based on the energy functional variational principle, the differential equation for beam vibration can be transformed into its weak form, reaching the integral expression for the rail's strain energy as follows:

$$\begin{aligned} U_B = & \frac{E_B I_B}{2}\int_0^{l_{uc}}\left(\frac{d\gamma}{dx'}\right)^H\left(\frac{d\gamma}{dx'}\right)dx' + \frac{\kappa_B G_B A_B}{2}\int_0^{l_{uc}} \\ & \times\left[\left(\frac{dv}{dx'}\right)^H\left(\frac{dv}{dx'}\right)-\left(\frac{dv}{dx'}\right)^H\gamma-\gamma^H\frac{dv}{dx'}+\gamma^H\gamma\right]dx', \end{aligned} \tag{2}$$

The symbol H in superscript position stands for the complex conjugate transpose. Employing the conventional energy integral formula yields a complex number as the result, lacking practical physical meaning. Thus, in this study, we employ the conjugate integral formula. The energy derived from this formula represents its amplitude, facilitating problem-solving. An example of such an application is in solving the structural characteristic equation, as demonstrated below. The Bloch theorem is employed in conjunction with the plane wave series expansion method. This approach leads to the derivation of a displacement expression for the beam, satisfying both the periodic boundary [15, 16] condition and the wave propagation condition. Therefore, the beam's displacement can be expressed as a Fourier series expansion, each of which represents a traveling wave component. According to the requirement of convergence, the infinite Fourier series can be truncated, and the beam's displacement can be obtained as:

$$\begin{cases} v(x') = e^{ikx'} \sum_{j=-a}^{a} v_j e^{i\frac{2\pi}{l_{uc}}jx'} = \{\xi(x')\} \cdot \{\mathbf{A_1}\}, \\ \gamma(x') = e^{ikx'} \sum_{j=-a}^{a} \gamma_j e^{i\frac{2\pi}{l_{uc}}jx'} = \{\xi(x')\} \cdot \{\mathbf{A_2}\}, \end{cases} \tag{3}$$

In the truncated expansion for both $v(x')$ and $\gamma(x')$, the total number of elements is set to $(2a+1)$; the imaginary unit is denoted by i; $\{\xi(x')\}$ denotes the trial function row vector; k is the wave number; $\{\mathbf{A_1}\}$ and $\{\mathbf{A_2}\}$ stand for the column vectors compromising the Fourier coefficients. $\{\xi(x')\}$, $\{\mathbf{A_1}\}$, and $\{\mathbf{A_2}\}$ are as shown below:

$$\begin{aligned} \{\xi(x')\} &= \left\{ e^{ikx'+i\frac{2\pi x'\cdot(-a)}{l_{uc}}}, e^{ikx'+i\frac{2\pi x'\cdot(-a+1)}{l_{uc}}}, \cdots, e^{ikx'+i\frac{2\pi x'\cdot a}{l_{uc}}} \right\}; \\ \{\mathbf{A_1}\} &= \{v_{-a}, v_{-a+1}, \cdots, v_a\}^T; \\ \{\mathbf{A_2}\} &= \{\gamma_{-a}, \gamma_{-a+1}, \cdots, \gamma_a\}^T; \end{aligned} \tag{4}$$

The beam's displacement fields meet the periodic condition. Consequently, the internal force, including the beam's shear force, which is expressed as follows, also conforms to the periodic conditions:

$$Q_B = \kappa_B G_B A_B \left(\frac{dv}{dx'} - \gamma \right) = \kappa_B G_B A_B \left\{ e^{ikx'} \sum_{j=-a}^{a} \left[v_j \left(ik + i\frac{2\pi}{l_{uc}} j \right) - \gamma_j \right] e^{i\frac{2\pi}{l_{uc}}jx'} \right\} \tag{5}$$

where $e^{ikx'}$ and $e^{i\frac{2\pi}{l_{uc}}jx'}$ endow Q_B with the conditions of propagation and periodicity. Upon substituting Eq. (3) into Eq. (2), the result is obtained:

$$U_B = \frac{1}{2} \{\mathbf{d}_1\}^H [\mathbf{K_B}] \{\mathbf{d}_1\}, \tag{6}$$

where $\{\mathbf{d}_1\}^H = \left\{ \{\mathbf{A_1}\}^H, \{\mathbf{A_2}\}^H \right\}$. The symbol H in superscript position stands for the complex conjugate transpose. $[\mathbf{K_B}]$ is the beam's stiffness matrix.

So the beam's kinetic energy expression is shown as follows:

$$\begin{aligned} T_B &= \frac{\omega^2}{2} \int_0^{l_{uc}} \rho_B A_B v^2 v dx' + \frac{\omega^2}{2} \int_0^{l_{uc}} \rho_B I_B \gamma^H \gamma dx' \\ &= \frac{\omega^2}{2} \{\mathbf{d}_1\}^H [\mathbf{M}_B] \{\mathbf{d}_1\}, \end{aligned} \tag{7}$$

where $[\mathbf{M}_B]$ is the beam's mass matrix.

2.2.2 The functional form of the plate's energy

In this chapter, the Mindlin plate theory is used to replace the thin plate theory to describe the model, enhancing the proposed method's applicability. Therefore, the energy functional for Mindlin plate can be derived using SGM. Drawing on Mindlin plate theory, the vertical motion of a plate can be expressed as [17]:

$$\begin{cases} \dfrac{\partial^2 w}{\partial x^2}+\dfrac{\partial^2 w}{\partial y^2}-\dfrac{\partial \theta_x}{\partial x}-\dfrac{\partial \theta_y}{\partial y}+\dfrac{\rho_P\omega^2 w}{\kappa_P G_P}+\dfrac{\kappa_S}{\kappa_P G_P h}=0, \\ \dfrac{\partial^2 \theta_x}{\partial x^2}+\dfrac{1-\mu_2}{2}\dfrac{\partial^2 \theta_x}{\partial y^2}+\dfrac{1+\mu_2}{2}\dfrac{\partial^2 \theta_y}{\partial x \partial y}+\dfrac{\kappa_P G_P h}{D_P}\left(\dfrac{\partial w}{\partial x}-\theta_x\right)+\dfrac{\rho_P I_P \omega^2 \theta_x}{D_P}=0, \\ \dfrac{\partial^2 \theta_y}{\partial y^2}+\dfrac{1-\mu_2}{2}\dfrac{\partial^2 \theta_y}{\partial x^2}+\dfrac{1+\mu_2}{2}\dfrac{\partial^2 \theta_x}{\partial x \partial y}+\dfrac{\kappa_P G_P h}{D_P}\left(\dfrac{\partial w}{\partial y}-\theta_y\right)+\dfrac{\rho_P I_P \omega^2 \theta_y}{D_P}=0, \end{cases} \tag{8}$$

where θ_x and θ_y indicate the plate's angular displacements along the x and y axes; w is the plate's vertical displacement; the plate's bending stiffness is denoted by D_P, which is denoted by $E_P h^3/\left[12\left(1-\mu_P^2\right)\right]$. The Poisson's ratio is denoted by μ_P, Young's modulus is denoted by E_P, and the thickness is denoted by h; the plate's shear modulus is denoted by G_P, which is equal to $E_P/[12(1+\mu_P)]$; the plate's shear coefficient is denoted by κ_P; the plate's density is denoted by ρ_P; the plate's moment of inertia is denoted by I_P.

The governing equation from Mindlin plate theory suggests that the plate's strain energy can be expressed in a double integral. The concrete form is as follows:

$$\begin{aligned} U_P=&\frac{1}{2}D_P\iint\left[\left(\frac{\partial\theta_x}{\partial x}\right)^T\left(\frac{\partial\theta_x}{\partial x}\right)+\mu_P\left(\frac{\partial\theta_x}{\partial x}\right)^T\left(\frac{\partial\theta_y}{\partial y}\right)+2\mu_P\left(\frac{\partial\theta_y}{\partial y}\right)^T\left(\frac{\partial\theta_x}{\partial x}\right)+\left(\frac{\partial\theta_y}{\partial y}\right)^T\left(\frac{\partial\theta_y}{\partial y}\right)\right.\\ &+\frac{1-\mu_P}{2}\left(\frac{\partial\theta_x}{\partial y}\right)^T\left(\frac{\partial\theta_x}{\partial y}\right)+(1-\mu_P)\left(\frac{\partial\theta_x}{\partial y}\right)^T\left(\frac{\partial\theta_y}{\partial x}\right)\\ &\left.+\frac{1-\mu_P}{2}\left(\frac{\partial\theta_y}{\partial x}\right)^T\left(\frac{\partial\theta_y}{\partial x}\right)\right]dxdy+\frac{1}{2}\kappa_P G_P h\\ &\times\iint\left[\theta_x{}^T\theta_x+\theta_x{}^T\frac{\partial w}{\partial y}+\left(\frac{\partial w}{\partial x}\right)^T\theta_x+\left(\frac{\partial w}{\partial x}\right)^T\left(\frac{\partial w}{\partial x}\right)+\theta_y^T\theta_y\right.\\ &\left.+\theta_y^T\left(\frac{\partial w}{\partial y}\right)+\left(\frac{\partial w}{\partial y}\right)^T\theta_y+\left(\frac{\partial w}{\partial y}\right)^T\left(\frac{\partial w}{\partial y}\right)\right]dxdy \end{aligned} \tag{9}$$

Notably, even though the plate is a non-uniform periodic structure, it differs in the plate's distribution between the two models. In the beam-discrete model of the CRTS-III plate, the wave number k is exclusively featured in the beam's displacement field expansion term. It is absent from the plate's expansion term. In contrast, the beam-continuous model of the CRTS-II plate incorporates the wave number k in both the beam's and plate's displacement field expansion terms. Using SGM as a basis, the displacement fields for both plate types are expanded in the following ways:

$$\begin{cases} w = \sum\limits_{m=1}^{M}\sum\limits_{n=1}^{N} A_{mn}\varphi_m(x)\psi_n(y), \\ \theta_x = \sum\limits_{m=1}^{M}\sum\limits_{n=1}^{N} B_{mn}\varphi_m(x)\psi_n(y), \\ \theta_y = \sum\limits_{m=1}^{M}\sum\limits_{n=1}^{N} C_{mn}\varphi_m(x)\psi_n(y), \end{cases} \tag{10}$$

$$\begin{cases} w = e^{ikx} \sum\limits_{m=-M}^{M}\sum\limits_{n=1}^{N} A_{mn}e^{i\frac{2\pi}{l_{uc}}mx}\psi_n(y), \\ \theta_x = e^{ikx} \sum\limits_{m=-M}^{M}\sum\limits_{n=1}^{N} B_{mn}e^{i\frac{2\pi}{l_{uc}}mx}\psi_n(y), \\ \theta_y = e^{ikx} \sum\limits_{m=-M}^{M}\sum\limits_{n=1}^{N} C_{mn}e^{i\frac{2\pi}{l_{uc}}mx}\psi_n(y), \end{cases} \tag{11}$$

where Eqs. (10) and (11) offer the displacement field equations for the discrete and continuous plate models; A_{mn}, B_{mn}, and C_{mn} are unknown coefficients. Mutual orthogonality in the x and y directions is exhibited by polynomials $\varphi_m(x)$ and $\psi_n(y)$.

Traditional Fourier series, composed of cosine or sine series, pose challenges in guaranteeing the continuity and differentiability of their higher-order derivatives. To improve the high-order continuity and derivability, we use both cosine series and sine series to express the Fourier series. It has been demonstrated that these series possess a differentiable third-order derivative and a continuous fourth-order derivative. This enables them to meet general perimeter conditions. Consequently, the modified Fourier series is taken as the spectral function. This choice allows $\varphi_m(x)$ and $\psi_n(y)$ to be formulated as follows [18]:

$$\begin{cases} \varphi_m(x) = \sin(m\pi x), & 0 < m < 5, \\ \varphi_m(x) = \cos[(m-5)\pi x], & m \geq 5, \\ \psi_n(y) = \sin(n\pi y), & 0 < n < 5, \\ \psi_n(y) = \cos[(n-5)\pi y], & n \geq 5, \end{cases} \tag{12}$$

where $m = 1, 2, 3 \dots M$; $n = 1, 2, 3 \dots N$. The value of 5 is meticulously chosen to guarantee the continuity of the third-order derivative of the plate's trial function, and fourth-order derivative existence in the entire solution domain, overcoming the discontinuity that probably exists at the boundary effectively.

Eqs. (11) and (12) can also be expressed as follows:

$$\begin{Bmatrix} w \\ \theta_x \\ \theta_y \end{Bmatrix} = \begin{bmatrix} \{\lambda\} & & \\ & \{\lambda\} & \\ & & \{\lambda\} \end{bmatrix} \begin{Bmatrix} \{\mathbf{A}_0\} \\ \{\mathbf{B}_0\} \\ \{\mathbf{C}_0\} \end{Bmatrix}, \tag{13}$$

where $\{\mathbf{A}_0\}$, $\{\mathbf{B}_0\}$, and $\{\mathbf{C}_0\}$ are each uniquely characterized as N-dimensional column vectors, which can be formulated as

$$\begin{aligned}\{\mathbf{A}_0\} &= \{\mathbf{A}_{11}, \mathbf{A}_{12}, \cdots, \mathbf{A}_{mn}\}^T, \{\mathbf{B}_0\} = \{\mathbf{B}_{11}, \mathbf{B}_{12}, \cdots, \mathbf{B}_{mn}\}^T,\\ \{\mathbf{C}_0\} &= \{\mathbf{C}_{11}, \mathbf{C}_{12}, \cdots, \mathbf{C}_{mn}\}^T;\\ \{\lambda\} &= \{\varphi_1(x)\psi_1(y), \varphi_2(x)\psi_2(y), \cdots, \varphi_M(x)\psi_1(y), \cdots, \varphi_M(x)\psi_N(y)\}\\ &= \{Q_1(x,y), Q_2(x,y), \cdots, Q_{MN}(x,y)\},\end{aligned} \tag{14}$$

and for any elements among these, $\varphi_m(x)\psi_n(y) = Q_{a_2}(x,y)$, and $a_2 = (m-1)N + n$, where m is in the range of 1 to M and n is in the range of 1 to N. Nevertheless, when the plate has continuity, the vector λ should be expressed as shown below

$$\begin{aligned}\{\lambda\} &= \left\{e^{ikx+i\frac{2\pi}{l_{uc}}(-M)x}\psi_1(y), e^{ikx+i\frac{2\pi}{l_{uc}}(-M+1)x}\psi_2(y), \cdots, e^{ikx+i\frac{2\pi}{l_{uc}}Mx}\psi_1(y), \cdots, e^{ikx+i\frac{2\pi}{l_{uc}}Mx}\psi_N(y)\right\}\\ &= \left\{Q_1(x,y), Q_2(x,y), \cdots, Q_{(2M+1)N}(x,y)\right\},\end{aligned} \tag{15}$$

and for any elements among these, $e^{ikx+i\frac{2\pi}{l_{uc}}mx}\psi_n(y) = Q_{a_2}(x,y)$.
Substituting Eq. (13) into Eq. (9) yields:

$$U_P = \frac{1}{2}\{\mathbf{d}_2\}^T\{\mathbf{K}_\mathbf{P}\}\{\mathbf{d}_2\}, \tag{16}$$

where $\{\mathbf{d}_2\}^T = \left\{\{A_0\}^T, \{B_0\}^T, \{C_0\}^T\right\}$.

$$\begin{aligned}T_P &= \frac{1}{2}\omega^2 h \iint \rho_P \left[w^T w + \frac{h^2}{12}\left(\theta_x^T\theta_x + \theta_y^T\theta_y\right)\right] dxdy\\ &= \frac{\omega^2}{2}\{\mathbf{d}_2\}^T\{\mathbf{M}_\mathbf{P}\}\{\mathbf{d}_2\},\end{aligned} \tag{17}$$

2.2.3 Potential energy at the plate's boundary

Here, the calculation of the symmetric boundary should be considered. All of the boundaries except periodic boundaries in **Figure 2** can be simulated by a line spring and an angle spring due to the SGM. **Table 1** displays the spring stiffness [19, 20] values corresponding to various classical boundary conditions. Taking into account boundary 1 of the flat plate as depicted in **Figure 2(a)**, the plate boundary's potential energy is as shown below:

$$\begin{aligned}U_1 &= \frac{1}{2}k_1\int_0^l w^T(x,y=0)w(x,y=0)dx + \frac{1}{2}k_{qy}\int_0^l {\theta_y}^T(x,y=0)\theta_y(x,y=0)dx\\ &= \frac{1}{2}\{\mathbf{d}_2\}^T\begin{bmatrix}[\mathbf{K}_{\mathbf{F}1}] & & \\ & [\mathbf{O}_2] & \\ & & [\mathbf{K}_{\mathbf{F}2}]\end{bmatrix}\{\mathbf{d}_2\},\end{aligned} \tag{18}$$

Boundary condition	Stiffness of line spring k_l(N m^{-2})	Stiffness of angel spring k_{qy}(N rad^{-1})
Simply supported	∞	0
Fixed	∞	∞
Free	0	0
Symmetry	0	∞
Anti-symmetry	∞	0

Table 1.
Spring stiffness value table.

In the formula, k_1 and k_{qy} denote the stiffness coefficients of the uniformly distributed line spring and the rotating angle spring, respectively. $[\mathbf{O_2}]$ represents a $(M \times N) \times (M \times N)$ zero matrix, while $[\mathbf{K_{F1}}]$ and $[\mathbf{K_{F2}}]$ are $(M \times N) \times (M \times N)$ matrices.

The determination of the potential energy at the remaining three boundaries is achieved through the application of the method previously outlined.

2.2.4 Elastic foundation's potential energy beneath the plate

Here, we consider the elastic foundation to provide uniform support to the plate, simplifying it as a uniform spring. Thus, we can use the double integral to represent the potential energy stored in the elastic foundation beneath the plate, as detailed below:

$$\begin{aligned} U_2 &= \frac{1}{2}k_s \iint w^T w dx dy \\ &= \frac{1}{2}\{\mathbf{d_2}\}^T \begin{bmatrix} [\mathbf{K_{F1}}] & & \\ & [\mathbf{O_2}] & \\ & & [\mathbf{O_2}] \end{bmatrix} \{\mathbf{d_2}\}, \end{aligned} \tag{19}$$

The formula incorporates k_s, denoting the average stiffness of the elastic foundation beneath the plate [21]. Additionally, $[\mathbf{K_s}]$ represents a matrix of dimensions $(M \times N) \times (M \times N)$.

2.2.5 Potential energy of the spring at the beam-slab junction

It is connected to the beam through a wire spring. Thus, we can reach the representation of the spring's overall elastic potential energy in the model, as detailed below:

$$U_3 = \frac{1}{2}\int_0^l k_z(x')\left[v(x') - w(x, y_0)\right]^H \left[v(x') - w(x, y_0)\right] dx \tag{20}$$

where $k_z(x')$ is linking springs' vertical stiffness; and $\left[v(x') - w(x, y_0)\right]$ is the difference between the beam's displacements and the plate's, which is as follows:

$$v(x') - w(x, y) = \{\{\xi\}, \mathbf{O_{cv1}}, -\{\lambda\}, \mathbf{O_{cv2}}, \mathbf{O_{cv2}}\} \cdot \{\mathbf{A}\} \tag{21}$$

where $\{\mathbf{A}\}^H = \left\{\{\mathbf{A}_1\}^H, \{\mathbf{A}_2\}^H, \{\mathbf{A}_0\}^H, \{\mathbf{B}_0\}^H, \{\mathbf{C}_0\}^H\right\}$; $\{\mathbf{A}\}$ represents the unknown coefficient column vector. $\{\mathbf{A}\}^H$ stands for its vector of complex conjugate transpose; both $\mathbf{O}_{\mathbf{cv1}}$ and $\mathbf{O}_{\mathbf{cv2}}$ are 0-row vectors.

Substituting Eq. (22) into Eq. (21), the following equation results:

$$U_3 = \frac{1}{2}\{\mathbf{A}\}^H \begin{bmatrix} [\delta_{11}] & [\mathbf{O}_1] & [\delta_{13}] & & \\ [\mathbf{O}_1] & [\mathbf{O}_1] & [\mathbf{O}_3] & & \\ [\delta_{31}] & [\mathbf{O}_4] & [\delta_{33}] & & \\ & & & [\mathbf{O}_2] & \\ & & & & [\mathbf{O}_2] \end{bmatrix} \{\mathbf{A}\}, \tag{22}$$

where $[\delta_{11}]$, $[\delta_{13}]$, $[\delta_{31}]$, and $[\delta_{33}]$ are $(2a+1) \times (2a+1)$, $(2a+1) \times (M \times N)$, $(M \times N) \times (2a+1)$, and $(M \times N) \times (M \times N)$ dimensional matrices, respectively; the matrices denoted by $[\mathbf{O}_3]$ and $[\mathbf{O}_4]$ are, respectively, $(2a+1) \times (M \times N)$ and $(M \times N) \times (2a+1)$ null matrices.

2.2.6 Total energy functional

Having identified the energy profiles of each component part in the beam-slab coupling system in the previous sections, the total energy functional is as shown below:

$$\Pi = U_B + U_P + U_1 + U_2 + U_3 - T_B - T_P, \tag{23}$$

By taking the total energy function's derivatives concerning the unknown coefficients in Eq. (23), we arrive at the following expressions:

$$\frac{\partial \Pi}{\partial v_j} = \frac{\partial \Pi}{\partial \gamma_j} = \frac{\partial \Pi}{\partial A_{mn}} = \frac{\partial \Pi}{\partial B_{mn}} = \frac{\partial \Pi}{\partial C_{mn}} = 0 \tag{24}$$

Hereby, the structural vibration problem is simplified as an eigenvalue problem, and we can reach the equation:

$$\left([\mathbf{K}] - \omega^2[\mathbf{M}]\right)\{\mathbf{A}\} = \{\mathbf{O}\}, \tag{25}$$

The system's total stiffness matrix and mass matrix are denoted by $[\mathbf{K}]$ and $[\mathbf{M}]$, respectively.

For 1-D periodic structures [22–24], the first Brillouin zone is $[-\pi/l, \pi/l]$, and Eq. (22) can be used to derive the dispersion curve in this region. Briefly analyzing and manipulating the dispersion curve yields the band gap. Then, the formula model is employed to resolve the vertical vibration band gap calculations for the CRTS-III and CRTS-II track structures.

3. Numerical analysis

To assess the mixed solution's reliability, the proposed method is used in the study related to railway track dynamic behavior. Firstly, the convergence analysis confirms the calculation results' reliability. The influence of the three truncation terms, M, N, and a, as detailed in Section 2.2, on the band gap frequency is subjected to thorough analysis. Afterward, calculations are carried out for the vertical vibration band gaps related to the CRTS-II and CRTS-III track systems.

3.1 The application of the method to the CRTS-III track

3.1.1 Introduction

In China, CRTS-III slab ballastless track (for simplicity, the full text referred to as CRTS-III track) commonly follows a uniform basic unit design and is arranged longitudinally along the road. As shown in **Figure 3**, the basic unit is mostly formed of a self-compacting concrete (SCC) layer, rail, rail plate, and fastening system. The function of the SCC layer and the fastening system is to ensure the track structure's elasticity, which can be conceptualized as equivalent to connecting springs. Further, the track can be regarded as a pre-embedded beam structure. The adjacent slabs are spaced and autonomous, forming what are known as slab joints. Hence, the CRTS-III track exhibits a characteristic structure. This structure is characterized by its periodically coupled beam and discrete plate arrangement.

By referencing the CRTS-III track's structural features, we can construct a vertical vibration analysis model for its unit. This model is displayed in **Figure 4**. The orbit is assumed as a Timoshenko beam, incorporating periodic boundary conditions. The track slab, regarded as a Mindlin plate, has gaps between adjacent slabs. Of the four boundaries of the rail plate, one is symmetrical, while the other three are free boundaries. The SCC layer and fastening system are simplified as vertical support springs. The track plate is connected with the rail by fastening spring, and the foundation is linked to the track plate via the SCC layer spring.

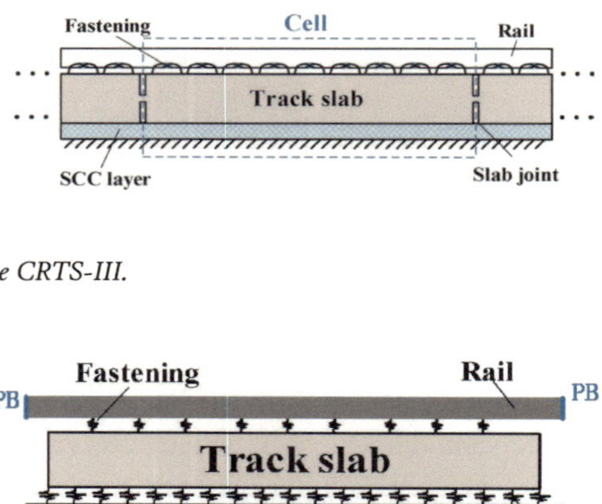

Figure 3.
The longitudinal section of the CRTS-III.

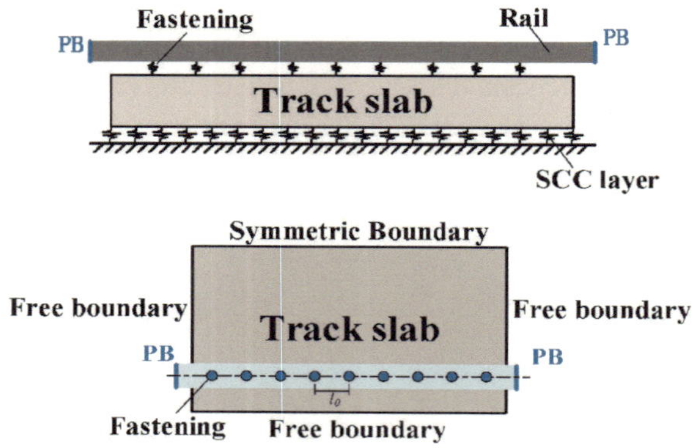

Figure 4.
The unit cell's model of the CRTS-III.

Mass density of the rail ρ_B(kg m^{-3})	**Cross-sectional area of the rail A_B(cm^2)**	**Young's modulus of the rail E_B(GPa)**
7850	77.45	210
Poisson's ratio of the rail μ_1	Fastening space l_0(m)	Length of the track slab l(m)
0.3	0.63	5.6
Thickness of the track slab h(m)	Young's modulus of the track slab E_P(GPa)	Poisson's ratio of the track slab μ_2
0.21	25	0.2
Width of the slab joint s_0(m)	Mass density of the track slab ρ_P(kg m^{-3})	Distance between the rail and FB y_0(m)
0.07	2300	0.4922
Area moment of inertia of the rail I_B(cm^4)	Half of the width of the track slab c(m)	Vertical fastening stiffness k_z(N m^{-1})
3217	1.25	7.2e7
Uniform support stiffness of the SCC layer k_d(N m^{-3})		
1.723e8		

Table 2.
Parameters related to the CRTS-III track.

It is crucial to recognize that the damping effect is not accounted for in the current model. Damping significantly affects the track structure's vibration response. This is particularly evident in the vibration response spectrum curve's peak value. Nevertheless, its influence on the structural characteristic frequency is relatively minor. That is to say, it exerts negligible influence on the track's band gap characteristics. Consequently, the influence of damping is usually dismissed in several theoretical studies on the properties of band gap orbits. **Table 2** outlines the relevant parameters of the theoretical model [25–28].

3.1.2 Vertical vibration dispersion curve analysis

In the existing research on the ballastless track's vibration characteristics, the track slab's influence is ignored or the track slab is considered as a beam. Few studies have modeled it as a slab. Therefore, before applying the method of this paper, it is essential to simplify the track slab into a finite-length mass block with infinite stiffness, ignoring the deformation of the track slab. This revised model should then be used to analyze the vertical vibration dispersion characteristics. The results are used as a control group to explore the superiority of the analysis method considering the track plate as a Mindlin plate in this chapter. With the aim of simplifying the analytical and computational process, the parameter m is introduced as a replacement for the wave vector k, satisfying the following condition:

$$m = \frac{kl}{\pi}, \tag{26}$$

In the above formula, k represents the wave vector restricted in the first irreducible Brillouin region.

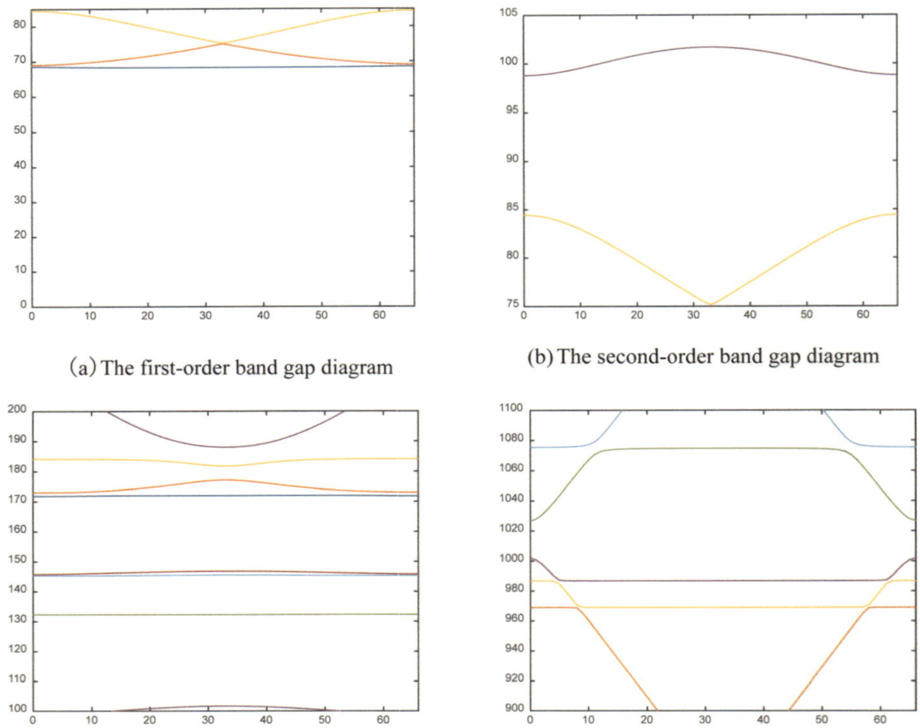

(a) The first-order band gap diagram

(b) The second-order band gap diagram

(c) The third-order band gap diagram

(d) The fourth-order band gap diagram

Figure 5.
The band gaps of the CRTS-III.

The orbital structure exhibits 1-D periodicity. This characteristic is most pronounced along the longitudinal direction of the orbit. For the 1-D periodic structure, the first irreducible Brillouin range, that is, the range of k is $[-\pi/l, \pi/l]$. The track structure's vertical vibration dispersion curve is illustrated in **Figure 5**. This curve represents the frequency band from 0 to 1200 Hz. The track structure's vertical vibration will produce a fourth-order band gap. The first three orders occur at low frequencies, whereas the fourth order appears at higher frequencies. In this context, the first-order band gap is formed through the transfer of energy to the soil.

3.2 The application of the method to the CRTS-II track

3.2.1 Introduction

CRTS-II slab ballastless track (the full text referred to as CRTS-II track) is a frequently used ballastless track type in China's high-speed rail infrastructure. Even though the track slab is manufactured in advance in a factory, the slab joint is poured with micro-expansive concrete, and the longitudinal connection of each slab is longitudinally reinforced. Therefore, the track slab can be considered to be continuously deployed along the line. Furthermore, the track structure can also be conceptualized as an infinite system comprising the track slab, rail, fastener, and layer of cement-emulsified asphalt mortar for adjustment, which is briefly referred to as the CA layer. As illustrated in **Figure 6**, the track structure's longitudinal section is depicted. The CA layer and fasteners endow the track structure with elasticity. Therefore, we can regard them as connecting springs. According to the information mentioned above, we can model the CRTS-II track as a periodic coupling of a beam and a continuous plate structure.

Similar to the prior part, we assume that the track slab and the rail are Mindlin plates and Timoshenko beams, respectively. The fastener is modeled as a linear spring for simplicity. And CA layer is depicted as a linear spring unit. Considering the symmetrical distribution of the central axis along the track structure, the track is modeled as a semi-structural model. This approach is taken to further reduce the computational load. Due to the periodic nature of the CRTS-II track, we focus on a single structural unit cell. We can construct the vertical vibration analysis model for the CRTS-II orbit, as shown in **Figure 7**. The applicable theoretical model parameters are detailed in **Table 3**.

Because of the model's small length, specifically when $M = N = 8$ and $a = 2$, the calculation results show convergence, and the track structure's dispersion curve remains stable. Thus, in the subsequent calculation, the previous three values are used as cut-off values.

The rail plate and rail are all solid elements, and the CA layer and fastener are assumed to be connecting springs. The rail and the track plate, as well as the track plate and the foundation, are linked by fastening springs and CA layer springs, respectively. Moreover, the rail inner plate is bounded symmetrically, and both ends of the rail plate and the rail are subject to Floquet periodic boundary conditions. The unit type is set to language-quadric.

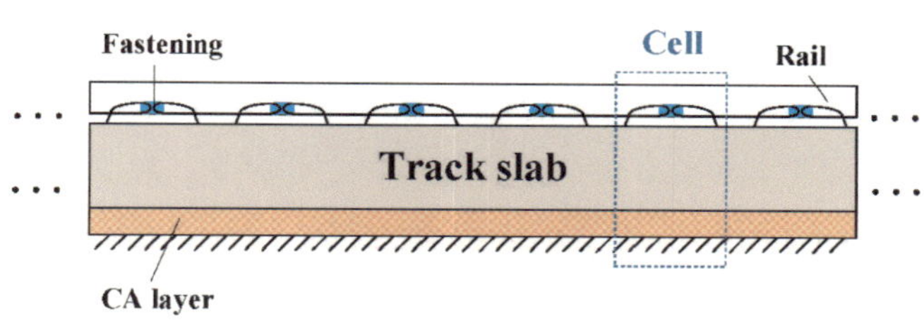

Figure 6.
The longitudinal section of the CRTS-II.

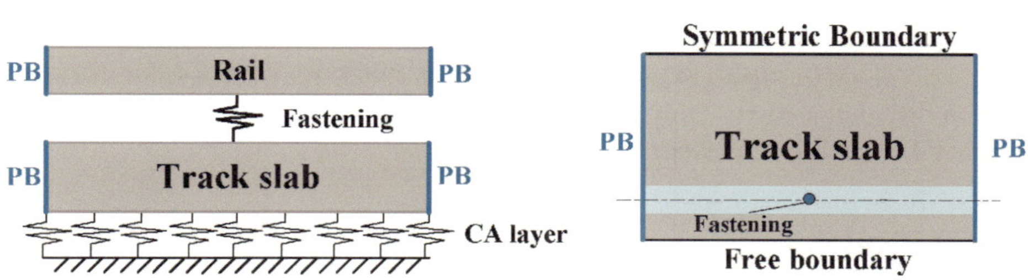

Figure 7.
The unit cell's model of the CRTS-II.

Mass density of the rail $\rho_B(kg\ m^{-3})$	**Cross-sectional area of the rail $A_B(cm^2)$**	**Young's modulus of the rail $E_B(GPa)$**
7850	77.45	210
Poisson's ratio of the rail μ_1	Mass density of the track slab $\rho_P(kg\ m^{-3})$	Length of the track cell $l(m)$
0.3	2300	0.63
Thickness of the track slab $h(m)$	Young's modulus of the track slab $E_P(GPa)$	Poisson's ratio of the track slab μ_2
0.21	25	0.2
Uniform support stiffness of the CA layer $k_d(N\ m^{-3})$	Fastening space $l_0(m)$	Distance between the rail and FB $y_0(m)$
1.723e8	0.63	0.4922
Area moment of inertia of the rail $I_B(cm^4)$	Half of the width of the track slab $c(m)$	Vertical fastening stiffness $k_z(N\ m^{-1})$
3217	1.25	7.5e7

Table 3.
Parameters related to the CRTS-II track.

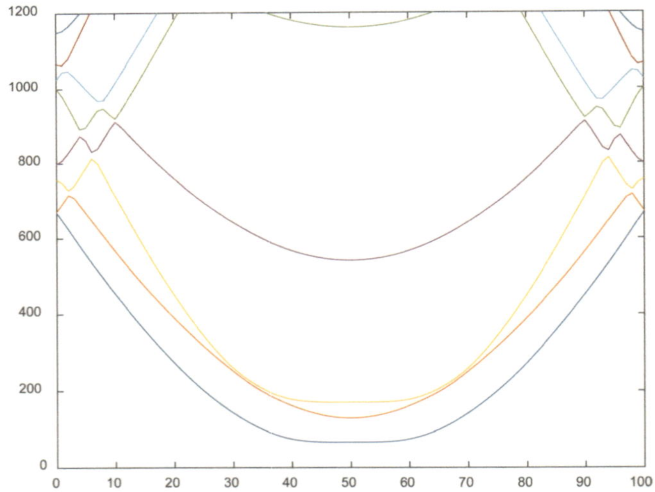

Figure 8.
Chart of dispersion curve of CRTS-II type plate.

3.2.2 Vertical vibration dispersion curve analysis

What is shown in **Figure 8** are the results of the CRTS-II orbital structure's vertical vibration dispersion curve. Only a vertical vibration band gap exists, spanning a frequency range of 0 to 65.9 Hz.

4. Conclusion(s)

The chapter applies phononic crystal theory to the study of vertical vibration band gaps. This approach is used for both periodically discrete tracks, such as the CRTS-III track, and continuous tracks, like the CRTS-II track. The discrete slab track, characterized by its continuous distribution within the CRTS-II track structure, exhibits a significantly greater number of band gaps compared to the continuous slab ballastless track. As a result, upon transmission from the rail to the slab via the fasteners, a greater number of elastic waves is able to continue propagating through the track slab. This leads to the elimination of certain vertical vibration band gaps. Considering the band gap characteristics enables the precise determination of the frequency range of elastic wave propagation within each orbital component, thereby providing a theoretical foundation for subsequent vibration mitigation efforts.

Author details

Wenjie Guo
East China Jiaotong University, Nanchang City, China

*Address all correspondence to: guowenjie@ecjtu.edu.cn

References

[1] Muhammad TW, Lim CW. Forest trees as naturally available seismic metamaterials: Low frequency Rayleigh wave with extremely wide bandgaps. International Journal of Structural Stability and Dynamics. 2020;**20**(14): 2043014

[2] Nobrega D, Gautier F, Pelat A, et al. Vibration band gaps for elastic metamaterial rods using wave finite element method. Mechanical Systems and Signal Processing. 2016;**79**:192-202

[3] Miranda EJP Jr, Nobrega ED, Ferreira AHR, et al. Flexural wave band gaps in a multi-resonator elastic metamaterial plate using Kirchhoff-love theory. Mechanical Systems and Signal Processing. 2019;**116**:480-504

[4] Sheng X, Zhong T, Li Y. Vibration and sound radiation of slab high-speed railway tracks subject to a moving harmonic load. Journal of Sound and Vibration. 2017;**395**:160-186

[5] Qian D, Wu J, He F. Electro-mechanical coupling band gaps of a piezoelectric phononic crystal Timoshenko nanobeam with surface effects. Ultrasonics. 2021;**109**:106225

[6] Qian D, Shi Z, Ning C, et al. Nonlinear bandgap properties in a nonlocal piezoelectric phononic crystal nanobeam. Physics Letters A. 2019; **383**(25):3101-3107

[7] Zhang R, Fu Q-X, Lu J-F. Analysis of the wave localization in a disordered periodic bridge with pile-soil-water-structure coupling. International Journal of Structural Stability and Dynamics. 2015;**15**(2):1450045

[8] Ashari AK, Stephen NG. On wave propagation in repetitive structures: Two forms of transfer matrix. Journal of Sound and Vibration. 2019;**439**:99-112

[9] Wang P, Yi Q, Zhao C, et al. Wave propagation in periodic track structures: Band-gap behaviours and formation mechanisms. Archive of Applied Mechanics. 2017;**87**(3):503-519

[10] Wang P, Yi Q, Zhao C, et al. Elastic wave propagation characteristics of periodic track structure in high-speed railway. Journal of Vibration and Control. 2019;**25**(3):517-528

[11] Feng Q, Dai C, Guo W, Yang Z, Lu J. A hybrid solution for band-gap analysis of vertical vibration for periodic beam-plate coupled systems based on variation principle. International Journal of Structural Stability and Dynamics. 2021; **21**(12):2150173

[12] Carter BG, McIver P. Water-wave propagation through an infinite array of floating structures. Journal of Engineering Mathematics. 2013;**81**(1):9-45

[13] Xiang H, Shi Z. Analysis of flexural vibration band gaps in periodic beams using differential quadrature method. Computers & Structures. 2009;**87**(23–24):1559-1566

[14] Xie X, Zheng H, Qu Y. A variational formulation for vibro-acoustic analysis of a panel backed by an irregularly-bounded cavity. Journal of Sound and Vibration. 2016;**373**:147-163

[15] Tang L, Cheng L. Broadband locally resonant band gaps in periodic beam structures with embedded acoustic black holes. Journal of Applied Physics. 2017; **121**(19):194901

[16] Deng J, Guasch O, Zheng L. A semi-analytical method for characterizing

vibrations in circular beams with embedded acoustic black holes. Journal of Sound and Vibration. 2020;**476**: 115307

[17] Mindlin RD. Influence of rotatory inertia and shear on flexural motions of isotropic, elastic plates. Journal of Applied Mechanics. 1951;**18**(1):31-38

[18] Ye T, Jin G, Shi S, et al. Three-dimensional free vibration analysis of thick cylindrical shells with general end conditions and resting on elastic foundations. International Journal of Mechanical Sciences. 2014;**84**:120-137

[19] Tang S, Han J. Simulation study of acoustic refraction wave manipulation based on sub-wavelength artificial periodic structure. Modern Physics Letters B. 2021;**35**(4):2150082

[20] Yang Y, Yau J, Urushadze S. Wave attenuation of a pre-tensioned wire with periodic spring supports subjected to a moving force. International Journal of Structural Stability and Dynamics. 2020; **20**(11):2071009

[21] Jin G, Zhang C, Ye T, et al. Band gap property analysis of periodic plate structures under general boundary conditions using spectral-dynamic stiffness method. Applied Acoustics. 2017;**121**:1-13

[22] Wang Z, Shen X, Wang S, et al. Subwavelength periodic structures design and its role in photon control of thin film solar cells. Nano. 2020;**15**(11): 2050139

[23] Tang L, Yao X, Wu G, et al. Band gaps characteristics analysis of periodic oscillator coupled damping beam. Materials. 2020;**13**(24):5748

[24] Sheng X, Li M. Propagation constants of railway tracks as a periodic structure. Journal of Sound and Vibration. 2007;**299**(4–5):1114-1123

[25] Chen M, Jin G, Zhang Y, et al. Three-dimensional vibration analysis of beams with axial functionally graded materials and variable thickness. Composite Structures. 2019;**207**:304-322

[26] Huang Z, Chen Z. Acoustic waves in two-dimensional phononic crystals with reticular geometric structures. Journal of Vibration and Acoustics-Transactions of the ASME. 2011;**133**(3):031011

[27] Zhou X, Wang Y, Zhang C. Effects of material parameters on elastic band gaps of two-dimensional solid phononic crystals. Journal of Applied Physics. 2009;**106**(1):014903

[28] Wang G, Wen X, Wen J, et al. Two-dimensional locally resonant phononic crystals with binary structures. Physical Review Letters. 2004;**93**(15):154302-154302